Evolution Family Reader

by

Eric F. Magnuson

Fimbul Winter Books

Evolution Family Reader

ISBN-13: 978-1512225723

ISBN-10: 151222572X

Evolution Family Reader

Evolution with Black and White
Pictures of Living Organisms
For Parents and Kids.

To

Cro-Magnon Man

and

Dr. Robert Webb

Table of Contents

Introduction   7

For the Family:

1, What Evolution Is   8

2. What Evolution Is Not   10

For Parents:

3. Matter and Evolution   11

4. Evolution and Devolution   18

5. Racial Displacement   26

6. Evolution and Revolution   38

7. The Triumph of Evolution   42

8. Living Organism Pictures   88

Free Books Online   200

About the Author   203

Fimbul Winter Books   204

Introduction

It is essential to the survival on Earth that the basic facts of evolution be understood by all people. The content here is divided into two parts. The first is for families and can be understood easily by all. The second is for parents, will need to be explained to children, and should be understood in the contexts from which it is taken, Evolutionary Psychology and Libertarian Nationalist Revolution by the current author. Special thanks to Dr. Hart Wegner of UNLV for helping to underscore the necessity of compiling this material.

Eric F. Magnuson

April 19, 2015
11:21 A.M.
UNLV

1. What Evolution Is

Evolution is the process by which living organisms adapt to their environment through change. This is necessary for survival, because the environment itself is always changing. If this happens so quickly that creatures can't keep up, they perish. Rapid environmental change usually has to do with temperature or water levels.

Genes are the cellular component within living organisms responsible for the characteristics of the organism. Genes themselves are continually changing. This process is called mutation and occurs randomly, but is sometimes triggered by environmental stimuli (e.g. increased solar activity).

When a creature's mutating genes result in a characteristic which favors adaptation to the environment, then the creature will live longer and reproduce more. The favorable genes are passed along to the offspring. If the opposite occurs, and the genes are not favorable to survival, then the creatures will die younger and reproduce fewer offspring, so the unfavorable genes are not passed on. This process is called natural selection, or survival of the fittest.

Science isn't science without proof. There are six proofs for evolution:

1. Universal Genetic Code

2. Continuity of Fossil Record

3. Interspecies Genetic Commonalities

4. Prenatal Growth Recaps Phylogeny

5. Postnatal Imprinting Recaps Phylogeny

5. Bacterial Resistance to Antibiotics

Note:

For detail on five of these proofs see Richard Peacock's site "Evolution: Frequently Asked Questions"

2. What Evolution is Not

Ever since Charles Darwin first explained the evolutionary process, there has been an unnecessary unwarranted feud raging between the religious and scientific communities. There is no real bone of contention here, and never has been, for three reasons:

1. Evolution is not a theory to be debated, but a proven scientific fact.

2. The fact that most observable phenomena are not mentioned in ancient scriptures, does not render them nonexistent.

3. Evolution does not negate the process off intelligent design. It is simply the means by which intelligent design is implemented. Universal intelligence is simply the potential for manifest existence residing in unmanifest existence (e.g. the light bulb before Edison). Natural selection unlocks this potential in the same manner as does an inventor. Both universal intelligence and technology are infinite, so there is a great deal to look forward to. See more about this ahead.

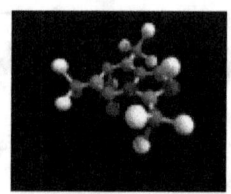

3. Matter and Evolution

1. Many who do not accept the fantasy world of popular religion will nonetheless conceptualize a bit beyond the easily provable to ideas about cosmic mind, universal intelligence, or the infinite. Preconceptions about the meaning of these terms should be put aside in this context.

2. Universal intelligence is not a religious construct involving belief. It is simply the potential for manifest existence residing in un-manifested existence. For example, an invention "exists" before the inventor brings it to manifestation. This principle underlies all existence and is infinite. The potential for expression of universal intelligence is also infinite. Technology is one form of this expression, unlocked but not created by man, and is itself infinite.

3. Universal memory, or true history, is the immutable fact of prior manifest existence. For example, an invention once did exist, and still "exists" even after it falls into non manifestation and is completely forgotten by all people (1).

4. Universal destiny, or futurity, is the immutable fact of as yet un-manifested existence. The only thing which can make the future better is constructive activity now.

5. Past, present, and future are in constant parallel occurrence. Our passage through time is merely experienced by us as sequential duration (2) . This time continuum principle and the possibility of time travel are, of course, General Relativity as outlined by Albert Einstein. In

this context time can be thought of simply as resembling an additional physical dimension.

6. Cosmic purpose manifests as universal intelligence finds ever more varied and complex expression through energy, in matter.

7. Nature is the actualization of universal purpose through the unimpeded functioning of universal laws. The laws of nature are universal laws.

8. Universal laws are the fixed relationships of action, reaction, and interaction which exist among spirit, energy, and matter in the universe as a whole.

9. Spirit is the expression of universal intelligence through good and evil. For example, while intellect will allow the individual to devise brilliant methods of hurting others, it is spirit which makes the individual question the rightness of using these methods.

10. Soul is that eternal archetypal essence of universal intelligence of which the individual spirit is the evolving and manifest representative in matter.

11. Energy is the means by which universal intelligence expresses itself through matter. This is accomplished via motion.

12. Matter is that which gives form to the expression of universal intelligence by energy. Matter is neutral from a moral standpoint and not evil as many have asserted.

13. There is the dichotomy of spirit and matter. Characteristic of spirit itself there is the dichotomy of good and evil, from a moral standpoint conceptually above and below matter respectively. Characteristic of matter itself are only varying degrees of utility as a vehicle for spirit.

15. In what is called an oscillating universe, matter spiraling out from Big Bang would eventually return to one point, then re-compacting to cause a new explosion. This can be explained as a simple process of temperature. All matter is attracted to all other matter by gravity and moves together accordingly. When molecules get too close the vibratory energy results in collision. The friction produces heat. When the heat excitation increases to a certain point, gravity is suddenly overcome and explosion outward occurs. Then begins the process of cooling. This allows the molecules to start moving back together forming stars and planets. Gravity determines orbits. Further cooling allows gravity to make orbits continually smaller. Eventually everything spirals back in together and a new explosion occurs (3).

16. Evolution begins when universal intelligence, through energy, moves matter towards the establishment of conditions which result in the existence of independently surviving, self-conscious, self-sustaining, self-perpetuating-organisms – life (4).

17. Even life, in all its complexity of manifestation, is at one level only a higher elaboration of the imprtus of matter towards motion.

18. Evolution continues as independently surviving organisms, through natural selection, give ever more varied and complex expression to universal intelligence.

19. Cosmogenesis, the realm of matter, is responsible for quantification and provides a vehicle for moral potential. Evolution or biogenesis, the realm of spirit, is responsible for qualitation and the actualization of moral potential.

20. The growth process of living things is dependent upon the ability to produce healthy new cells. The aging process leading to death occurs because the growth process is stopped by enzymes. One function of this is to limit the size of animals to accommodate increasing numbers and dwindling food supplies. Many species have gotten generally smaller since prehistoric times. This is an evolutionary adaptation. Death itself also has a useful function. Among species on land, if the growth process were to continue indefinitely, creatures would become so large that their weight would break their own skeletons. Some life forms in the sea lack this adaptation. They can slowly grow larger because water buoys them up and permits tremendous increase in weight. Barring fatal predation or environmental cataclysm, such creatures can live an extremely long time. When animals came out of the sea onto the land, the physical death of individuals served as a mechanism insuring survival of the species. Death is thus ironically an adaptation having "survival value" (5).

21. The closest man can ever come to actually creating anything is, through inspiration and effort, to discover pre-existing principles or energies and to arrange or apply these in previously un-manifested ways. The inventor essentially attunes himself with, and becomes a vessel for, the expression of universal intelligence.

22. Man is not the end product of evolution but an evolving expresser of universal intelligence. The potential for evolutionary expression is infinite because universal intelligence is infinite.

23. Evolutionary destiny is the imperative for the unimpeded, ever more varied and complex expression of universal intelligence through evolving organisms.

24. Since stars expend themselves and solar systems die, an uninterrupted progress for higher evolutionary expression depends upon the ability of life forms to either relocate, or at the very least, to communicate throughout the universe.

25. In an oscillating universe the progress of evolutionary expression would be destroyed cyclically. This, however, should not be taken as an excuse for thinking in the short term but would render any human interference with evolutionary destiny an even greater encroachment than otherwise.

26. Even if Big Bang explosions followed by re-compacting take place at different times and different places throughout the universe, then an infinitely uninterrupted future for evolutionary expression would still also depend, at very least, upon communication throughout the universe.

Footnotes:

1. In the East this principle is referred to as the Akashic Record.

2. Just as the Earth is subjectively experienced by us as being flat and stationary with the Sun moving around it.

3. This eternal process of the oscillating universe reminds us of what the ancient Hindus described as "the exhaling and inhaling of the cosmic breath," what the Old Norse called "Ragnarok," or more recently, what the Pagan scientist Hans Horbiger designated as "the eternal cosmic battle between fire and ice."

4. It has been proposed that the old question of which came first, the chicken or the egg, is somehow a great dilemma. Obviously the egg came first, because that mutant or husbanded life form which man would designate a chicken, resided within an egg laid by a mother who was sufficiently unlike her offspring so that man would not have designated her a chicken.

5. The idea of death having "survival value" is from Guy Murchie, p. 186, Old Farmer's Almanac 1977. This one phrase triggered the rest here. Self-preservation, of course, is a stronger instinct than preservation of the species. As we proceed up the phylogenetic ladder creatures increasingly determine the direction of their own evolution. All of his this strongly suggests the utility of reincarnation. Alternative explanations of apparent reincarnation phenomena observed under hypnotic regression are invariably more far-fetched that the notion of reincarnation itself. Consider the case of the man in India who claimed he had reached the point of awareness from one incarnation to the next. As he lay dying he told his wife that he would soon be reborn, where, when, and to whom. The specified mother became pregnant and gave birth. The minute the new infant learned to speak he identified himself as the man who had died, where, when,

and with whom. The ultimate implication here is that to execute evil people is only to give them a new body. They will have to go through potty training and elementary school all over again, but soon we will have them back again with no increase in their soul wisdom. It has been said that the best remedy for evil people is to simply let them go on living, doing useful work in a environment favoring contemplation, where they can't hurt anybody else.

4. Evolution and Devolution

So that we are not accused of overkill, let us acknowledge in advance that we know that the superior viability of Capitalism renders any arguments regarding human evolution superfluous and unnecessary. We simply like to illustrate the connection between natural and economic laws.

1. No species, no matter how highly evolved, can go against nature in the long term without bringing about it's own destruction. Man, to survive, must arduously study the laws of nature, because he is the only species on Earth possessing the ability to temporarily violate these laws.

2. Some who have raised a lion and a lamb to lay down with each other have claimed that this is the natural way for these creatures. Of course it is, when people are giving the lion three big meals a day.

3. The truth seeker will reason deductively from universal principals to their proper application in specific instances. Those not at one with truth will usually reason backwards inductively from exceptional particular instances to false general premises.

4. The laws which govern human interaction are interwoven with the laws of the ecosphere, are every bit as mysterious, and like the latter, are far too complex to permit regulation of the consequences which proceed from them. Indeed, the

affairs of men and of nature are one, and any attempt to unnecessarily regulate either is dangerous. One breech of natural law will produce a distortion which seems to call for yet another breech, and so on. The only way things can be "regulated" in a way consistent with survival on this planet is through a policy of non-interference with nature in the first place.

5. Man is an animal to be sure, but he is also something a bit more than just that. He is sometimes able to admit ignorance and, by his own volition, to develop himself intellectually and spiritually. By contrast, he also possesses the one characteristic which will allow him to deceive himself enough to destroy, first his liberty, then his environment, then himself, and that characteristic is rhetoric.

6. Rhetoric, in contrast to truth, is often used by the average individual to quickly rationalize the "rightness" of inaction relative to involvement in sweeping social change. Action in such areas might of course, interfere with activities of a frivolous nature, or even with pursuits of truly great and enduring human significance like ball games.

7. Congenital traits in human beings result from one of four known causes: heredity, random genetic mutation, intra-uterine conditions, or events during birth.

8. Notice how we refer to a quiet, serene baby as "good," and to a baby who cries all the time, as "bad." These casual labels reveal deeper truths.

9. Environment does not play the part in determining human behavior which many would like to believe. Intrinsic good moral character will often resist and win out against immoral upbringing in the long term. Sometimes good moral character will even temporarily over react, as in the case of the approach-avoidance conflict, where a cowardly parent will produce a war mongering child or vice versa in the short term, until life teaches maturity and balance.

10. We can all think of people, who from early childhood, manifest true seriousness and wisdom. We can also think of people, who even in advanced age, manifest only frivolity and low self-importance. One already knows much from the time of birth. The other learns almost nothing from a lifetime of experience. In a multi-media culture where we are all bombarded with vast amounts of information, these individual differences do not reflect the influence of environment alone. Let us note that those who think that society can be remade simply by changing peoples' environment for one generation, usually disgrace themselves further by advocating fanatical and un-Libertarian methods.

11. Natural selection is the principle that organisms manifesting physiological or behavioral traits having little survival value, will reproduce at a lesser rate than other organisms since they will not live long enough to do so. This will ultimately cause the trait to die out in the species, unless the particular organisms manifesting the non-surviving traits are given artificial sustenance that will prolong their lives and periods of reproduction.

12. Among non-human species, non-self-sustaining individuals are those, which for any reason, are unable to behave in a way consistent with their own survival in the long term in the context of a natural ecology relatively undisturbed by man. Nature eliminates such creatures through natural selection, and allows evolution to progress. Evolutionary progress depends just as much upon the non-breeding of these individuals, as upon the breeding of self-sustaining individuals.

13. When non-self-sustaining organisms are sustained artificially and end up reproducing when they would not have otherwise, then evolution stops and devolution begins. Devolution is the reversal of evolution. The detriment to evolutionary destiny is even more greatly magnified if the evolutionary expression of self-sustaining organisms is encroached upon in the process.

14. Non-human species on Earth are evolving in directions different from man, and must be allowed to do so normally. Ironically, the primary danger to this occurs as the indiscriminate proliferation of humans begins to reflect man's interference with his own evolution. Allowing a non-human species to evolve normally does not entail finding a way to sustain the weak elements within that species, nor does it consist in murdering their natural predators because of the alleged "cruelty" with which these predators take their prey. It simply involves leaving them alone.

15. Predation of non-human species by man may strengthen or weaken a species, depending upon which members of the species are killed. This in turn, is often determined by the method of predation used. The outdoorsman or professional hunter should contemplate these matters deeply. Should not this ancestral joy in hunting also be available to future generations?

16. Among human beings, non-self-sustaining or unproductive individuals are those who lack the ability for whatever reason, to behave in a way which is consistent with their own survival in the long term under the conditions normally prevailing in a totally free society without having to make unjust encroachment upon other people or the environment.

17. Briefly stated, a self-sustaining or productive member of the human species is simply one who is capable of surviving in a perfectly free society without needing to make unjust incursion against the liberty of others.

18. Nature is able to systematically ascertain non self-sustaining members of animal species and to eliminate them through natural selection. Man is the only animal able to temporarily gainsay nature in this regard, because man is the only species whose unproductive members have ambitious champions possessing rhetoric.

19. It seems that among "nice" or "civilized" people, it is quietly and conveniently forgotten that evolution applies as much to humans as to any other species. Cowardly religions even falsely teach in many cases, that evolution is merely a theory. In actuality man has reached the point on the evolutionary ladder where he not only can, but to survive must, will the direction of his own evolution. This can only be accomplished by working with nature, not against it.

20. Mankind will not survive if societies continue to be subsidized breeding farms for human non self-sustainability. Recognition of this does not mean that anybody needs to go hungry or starve. In a Libertarian Republic all people will live long happy lives. Ongoing sustenance of chronically unemployable people, however, can be provided much more efficiently by means other than coercive government. It is also possible through education to help these people to greatly curb the encroachment they make by excess reproduction.

21. It is important to note that in a collectivist society the number of truly non-self-sustaining people is always far fewer than would appear on the surface. The economic cycle caused by fiat monetary policy pulls down a tremendous number of people to a level of apparent non-self-sustainability. By getting rid of this one simple ecomic policy alone, unemployment would thereafter not exceed one percent except in cases of local catastrophe.

22. It has been empirically proven that, through natural selection, humanity was able to evolve from less complex forms only because nature offers no sustenance to non-self-sustaining organisms. In man's earlier development he was more at the mercy of nature than today

23. Man has not been able to cheat nature in the longer term. One effect of a warm climate and advancing medical science has been to endlessly thwart natural selection. This has devolved the human species both physically and

intellectually. Cro-Magnon Man was not only bigger and stronger than *homo-sapiens*, but with an average IQ of 165, was a race of geniuses by comparison. The decline in intelligence which we enjoy today was produced by thirty thousand years of warm weather and easy living preceding the last Ice Age (19).

24. More intelligent people tend towards self-actualization at the expense of procreation. Less intelligent people are outbreeding more intelligent people everywhere (20). The diaries of the Roman Emperor-Historian Claudius show that this trend had become a matter of concern even in the time of Augustus, who spoke publicly about it. The problem is accelerating at a more rapid rate today. The excess proliferation of less intelligent humans must be seen as a direct encroachment against all life on this planet if any reasonable quality of life is to be preserved, or ultimately if humanity itself is to survive (21).

25. Because of the prevalence of collectivist ideas, and man's foolish presumption that he can ignore natural law, unproductive humans are protected from the consequences of their own actions. This allows them to reproduce at a rate which they would not be able to sustain ordinarily. Such people see a large number of children as their insurance policy for old age.

26. Experience has shown that when people are prosperous because of their own activity, as under a Libertarian Capitalist system, that they will restrict their numbers by their own volition. Non-self-sustaining individuals cannot survive without aid in the short term, or even in the long term, because they will indiscriminately increase their numbers to the point where even their protectors can no longer help them. When this happens, famine will cause far more suffering than would have been the case without protectionist interference in the first place.

27. Legislative measures to restrict birth rates among people at lower achievement levels have been proposed, are well thought out, and are technically workable (22). Of course, if we had Libertarian Capitalist economic systems in place plus education in the schools about the problem of human devolution, such measures would never become necessary. If we are going to do something, however, we had better get on with it, because in just forty more years the population of Earth is going to double.

28. The highest of men on Earth are as much above the lowest, intellectually and spiritually, as the lowest are above the chimpanzees. Those who don't instinctively know this identify themselves as being closer to the low end of the continuum.

29. What an individual learns in a lifetime is to some extent, passed on genetically. The forward progress of evolutionary expression is impeded if productive individuals are kept from learning by being sacrificed for unproductive individuals. Those who work diligently towards the devolution of life on Earth are being masochistic on a cosmic scale. The premise that excellence and evolutionary expression should be even slightly impeded to help sustain mediocrity is the beginning of the worst mistake which humanity could ever make.

30. The intelligent and capable "champion" promoting unjust encroachment by unproductive humanity, is the worst among them, because his inability to behave in a way which is consistent with his own and our survival in the long term, comes not from stupidity, but from moral inferiority. The truth will never reach these individuals, because they simply lack the capacity to comprehend any concept as exalted as evolutionary destiny

31. It is a pity how vocal are the people who do not "believe" in the empirically proven fact of evolution, but who do endorse more or less evolutionary social policies. Mixing the ridiculous rhetoric of fanatical religiosity with Capitalist

economics brings a popular perception of discredit upon economic truth because of guilt by association.

Footnotes:

19. Elmer Pendell, "Why Civilizations Self Destruct" (Howard Allen, Cape Canaveral, 1977). Quite scholarly and well documented. Essential reading for all caring people on Earth.

20. Ibid.

21. If we believe the accounts of alien molestation, it would seem that more advanced peoples have evolved in intelligence, but have devolved physically. Apparently they use human ovi and semen to upgrade themselves physically before they perish from sheer puniness and are doing this without devolving their intelligence because they use genetic engineering rather than simple fertilization.

22. Ibid.

5. The Truth about Worldwide Racial Displacement

Intelligent Humanity is an Endangered Species

Evolutionary destiny is the imperative for the unimpeded, ever-more varied and complex expression of intelligence. The mechanism for this is natural selection, the principle that organisms manifesting traits having negative survival value, will reproduce at a lesser rate because they will not live long enough to do so. This will ultimately cause the trait to die out in that species, unless those manifesting the traits are given unnatural sustenance to prolong their lives and periods of reproduction.

When non self-sustaining organisms are sustained artificially and end up reproducing when they would otherwise have not, evolution stops and devolution begins. Devolution is the reverse of evolution. The damage to evolutionary destiny is even worse if the evolutionary expression of self-sustaining organisms is encroached upon in the process. These principles apply to all life forms, including humanity.

Evolution will not continue, and ultimately intelligent mankind will not survive, if societies continue to act as subsidized breeding farms for human non-self-sustainability.

Recognition of this does not mean that anybody needs to go hungry or starve. In a Libertarian Republic all people can live long happy lives. Ongoing sustenance of chronically unemployable people can be provided efficiently without dragging down human excellence using coercive government wealth redistribution. It is also possible with education about birth control to help people to curb the encroachment they make by having more than two children in a world that passed the ideal population of 320 million people in the year 900 A.D.

Race Preservation is Not Injustice

It's perfectly normal to feel more comfortable and at ease among one's own kind. This is an instinctive trait which comes from tens of thousands of years of fierce tribal competition for food and shelter. Nobody should ever allow themselves to be put on the defensive about having these normal feelings. Sometimes normalcy in this regard is mitigated by other things: spirituality, education, fashion, brainwashing, fear, lust, insanity, greed, naivety, social masochism, self-hatred, stupidity, or any combination of these.

It's safe to say that a person's feelings are always based upon the total of what they have experienced. Most so called prejudice is just a normal human reaction to what an individual has experienced. Since we all have different experiences, we all have different reactions.

A stereotype is the random generalization that an individual member of an identifiable group, for better or for worse, possesses a particular characteristic, which may or may not be more common among members of that group, this without any substantive knowledge of the person as an individual. Stereotypes are usually based upon race, nationality, sex, age, and creed.

There is a double standard about constructive racial pride. If a member of a less highly evolved race shows it, he is praised for cultural consciousness, If a member of a more highly evolved race shows it, he is scorned for divisiveness or bigotry.

Dictionary definition: "Racism is the belief that race is the sole determinant of character in the individual". There are very few people who believe anything like this. Everyday experience contradicts it. When people speak the deeper truth about race, however, it's usually shouted down with catcalls about hatred and bigotry. What this tells us is that if we want to be thought of as being loving and open-minded, we have to go along with social lies.

Ideological falsehood about group member potential is always reinforced using induction, that is, reasoning backwards from exceptional particular instances to a false general premise. Individuals must be judged individually. That doesn't mean, however, that we have to be deaf, dumb, and blind as to what is true about groups, or the adverse effect that one group can have upon another when it is wrongfully displaced from its rightful ancestral homeland. What hurts a group, hurts individuals.

Evolved Intelligence is being Lost Forever

Probably the best definition of a race traitor is "one who will throw down everything that has been gained through four and a half billion years of evolution to be fashionably liberal minded sounding at cocktail parties". The differences between racial groups are based on far more than mere physical appearance. More importantly, there are significant differences in intelligence. This has many consequences.

Science shows that the seat of higher moral deliberation is the cerebral cortex. The ability to conceptualize morally is tied to overall intelligence. Less highly evolved races not only have a lower average IQ, but also a lower average moral IQ. This leads to poor relations with other groups. All one needs to do is to search "crime by race" on the Internet. Statistics show these things very clearly. Ivory tower wishful thinkers will, of course, deny, or twist, every nuance of truth in this area.

The interbreeding of a more highly evolved race with one less highly evolved, results in a new race somewhere between the two original races, in humans, usually closer in manifestation to the lower. Once racial interbreeding gets started in any society, devolution and downfall has begun.

It's usually the lower members of the more highly evolved race who interbreed with the upper members of the less highly evolved race. As time passes, more and more intelligence disappears overall, and group differences become increasingly blurred. Those of mixed race usually identify with the lower half of their ancestry out of a need for self-justification, and are usually not accepted by members of the higher race. The most noble thing such a person can do for evolutionary destiny in general, is to identify the higher race, abstain from the gene pool, and adopt children.

The Role of Politics

Ambitious representatives of less highly evolved races will always champion the genetic intermixing of their people with the more highly evolved race. Why wouldn't they? They have everything to gain and nothing to lose.

Interracial marriage advocates are trying to obliterate race by making all people into one race. Dominant genetic traits are those which win out over long periods of interbreeding. Examples are brown eyes and black hair. Recessive genetic traits are the ones that lose out. These include light eye colors, such as green, blue, and hazel. Also gone forever will be light hair colors, such as blond, ash, auburn, and red.

If one race possesses mostly recessive genetic traits, then that race will be destroyed by race mixing. Self-respecting members of such a race will quite properly perceive all the breed-up-quick philosophy as a threat to the continued existence of their own race. They will feel that interracial dating websites are not examples of open-mindedness or cultural progress, but only of ignorance and societal decay, something effectively akin to an epidemic of fatal disease. The real issue is racial preservation, diversity rather than sameness, variety rather than monoculture.

International finance manipulates politicians and always promotes any massive government spending program that will allow them to lend money. This includes ongoing programs that implement unworkable Socialist policies. They escape the cultural impact, because they can afford to live or vacation anywhere on Earth. People have the natural right to grow up in a society among their own racial kinsmen and should not accept being coerced at gunpoint by cruel socialist slave masters into intermixing with sullen, angry racial outlanders, who have an in-your-face attitude and speak with a tone of blaming.

It's important to know that third-world people usually favor globalization, because it will allow them to prosper from social programs paid for by productive host populations. Globalist bankers know that countries with multitudes of immigrant third-worlders, if globalization comes to a ballot referendum, will be far more likely to relinquish sovereignty. This is the reason for all the new indigent faces in productive countries.

Solution: Make the Whole World Free

Science teaches that when two groups compete for the same ecological niche, the stronger will destroy the weaker. Distance between the groups, of course, eliminates the competition. A good non-human example is timber wolves and coyotes. Both are *canis lupus*, but they are also natural enemies. Both survive very nicely if they are at a distance from each other. Coyotes like to travel around a good deal, but the smart ones have learned not go within seventy-five miles of timber wolves.

Those who would rob you of your liberty, or threaten the existence or evolutionary destiny of your race, are your mortal enemies. There are two ways to deal with them. You can either have them at a distance or eliminate them completely, at very least by stopping their further reproduction. Which you choose should be determined only by your perception of possibility and cost.

It's normal for the people of a more highly evolved race to take up arms against those who seek to destroy their race, and with it the evolutionary destiny of mankind. There are, however, more peaceful solutions.

Populations displaced by coercion have the natural right to return to their ancestral homelands. This option can be made legal through the cooperation of governments everywhere. Worldwide liberty and capitalism will produce worldwide prosperity. This will allow populations who have been unjustly displaced to return to their ancestral homelands, because the newly prospering nations will be able to easily accommodate their return.

Displaced people of reproducing age who are realistic, responsible, and mature will want to participate in this great adventure. They can entrust their property to older friends and relatives who choose to remain in the host counties. Their property can then be sold at a time advantageous from the standpoint of market.

There is no substitute for the splendid integrity of living in a place where you are wanted, and where you don't have to blame, or thank, anyone but yourself, for how well you do.

Eric F. Magnuson
October 15, 2008
Late Morning

The following is an adaptation of a essay written by Dirk Aubrey Lokison in 1983 about racial displacement problems particular to the United States. In the original version, specific ethnic and racial groups were identified. The WLO's goal is not to hurt anyone's feelings, just to make a better world, so I took these specific references out. The who-does-it is gone, but the what-they-do remains. See if you can ID the various groups. I also made editorial changes for easy readability, but not for meaning or content. Those of European heritage will think of many culturally displaced individuals who they like or love. Personal affiliations, however, must not interfere with more important matters such as race preservation and evolutionary destiny. Goals must be prioritized. We can visit our friends overseas when they are happy and prospering back in their ancestral homelands. - EFM

Displaced Racial Populations in America

People of European ancestry are the founders of America as it is now constituted, and must not be hindered in the establishment of a totally free society by antithetical groups who never belonged here in the first place. There are many specific problems in the USA which accrue to the presence of displaced populations. There is, of course, in every group a percentage of decent people. In the groups that are hurting the United States however:

- Many are significantly less intelligent. Those who question the accuracy of intelligence tests need only look at the long record of almost total non-achievement by certain groups in both America and in their countries of origin. They are often loud, belligerent, and foul-mouthed. Their presence in America spoils the quality of life for people of European heritage.

- Many have out-bred their ability to feed themselves at home. Americans should not have to put up with vast throngs of third-worlders coming into our country to do the same thing here. Most of the invaders are tedious and uninspired, addicted to indiscriminate reproduction and cowardly medieval religion.

- Many are cruel, fanatical, and given to terrorism. The dangerous ones exist in sufficient percentage so that we shouldn't have to worry about which are which.

- Many are conniving monoculturalists who are openly aggressive towards the host populations of European ancestry. This parasitic element sponsored and now exploits the Socialism that is destroying America. We shouldn't have to put up with the vile dispiriting nihilism of these wandering internationalist blood suckers.

- Many are intelligent, quiet, polite, and hard working. There are, however, enough of them in their homelands. In the interest of our own preservation, we should not welcome vast numbers of them here with their cheerless similarity of appearance. It's simply a bad move genetically from the standpoint of human aesthetics and variety.

Social lies are always used by collectivist governments to justify non-viable policies. There are two big lies guiding American social policy. One pertains to who did what. Virtually all black people in the USA are descended from those who were brought to America as slaves. Most white people in America, however, are not descended from greedy southern plantation owners, and have inherited no culpability in any of this. The presence of black people in America is, for most white people, simply an unjust cultural and genetic encroachment. The other big lie is that all free people were unjustly enriched by the institution of slavery. The truth is that the only people who gained from slavery were the plantation owners. Everybody else was hurt because of what it did to the price of commodities and the labor market.

The slavery lies are used to promote so called "social justice" via socialist wealth redistribution. Affirmative Action programs deny education to more qualified people of higher intelligence. Colleges in the United States have been invaded by loud, motor-mouth jokers who disrupt the study of serious students. When anyone asks them to please quiet down, they often respond defiantly with name calling and threats of violence. This is especially true first semester, before many of them flunk out. Anybody who doubts any of this need only sit for a while in any college library of computer lab and listen to what goes on, and then see for themselves who is responsible. Try riding a city bus over time and then ask yourself why ninety nine percent of the trouble consistently comes from twelve percent of the people.

Non-Coercive Solutions

Large percentages within all displaced population groups are antagonistic to the interests of liberty loving Americans of European descent. Only small percentages have demonstrated any real comprehension of free-enterprise principles. Their continued insistence on the injustice of Affirmative Action and unnecessary social programs is the most concrete expression of this.

Admitting the truth of all these things does not mean that anyone should condone cruelty, injustice, or disrespectful behavior towards anyone else. Nor should it be used to refute the great accomplishments of exceptional individuals within any displaced group.

On the average, however, people of European ancestry have everything to lose and nothing to gain through further association with displaced populations. The terrible injustices of history cannot be rectified by the further injustice of perpetuating Socialism in America. Two wrongs don't make a right.

Our greatness as a nation can now only be realized by doing what is clearly most workable. It is in our interest to help displaced people get back home to whatever country they came from. Mass influx should be stopped immediately. Incompatible groups already in the United States must be allowed to return to their countries of origin.

He sent certain animals to tell men that he showed himself through the beast, and that from them, and from the stars and the sun and moon should man learn...

- Eagle Chief(Letakos-Lesa) Pawnee

So called "Native" Americans are actually first wave European immigrants who came here via the Alaskan Land Bridge fourteen thousand years ago. They are the only people who truly deserve to be in America, but there are not enough of them left to defend the borders of a country this size. Most of them are gone because of diseases brought by later European settlers. As Americans we owe the survivors a great deal.

We should give Native Americans back good tracts of government land to inhabit, plus all of the National Parks and

Forests, to be run by them for the use of all Americans. Their religion is much like the original spirituality of Europeans. Whenever possible they should be given the autonomy of a separate nation, while at the same time, remaining our military allies relative to any invasion from outside America.

Dirk Aubrey Lokison

1983

Updated Commentary

In all of human history, there is no example of any multi-cultural society having survived. It has always led to the downfall of the society, because only those who create a great culture are capable of sustaining it. What works are indigenous peoples, enjoying race and culture preservation in absolute liberty as separate sovereign nations competing in a free world market. Only this, can lead to world peace and prosperity. We are tired of being victimized, and of explaining the facts to militant sleepwalkers. Revolution is at hand! Good people will, by any means necessary, make natural order prevail on this planet. Those who oppose us, dig your graves. Winter is coming.

Eric F. Magnuson
February 23, 2016
11:05 A.M.

6. Evolution and Revolution

[Many propositions here will be found in "New World Order: Just Say No!" and VMF "Libertarian Basics".]

1. Liberty is the inalienable birthright of every living organism in the universe to manifest justly as an unimpeded participant in evolutionary destiny. This manifestation, to be both Libertarian and just, must not unnecessarily interfere with the evolutionary expression of any other living organism.

2. Perhaps the most noble and heroic trait of mankind is the innate impulse towards individual liberty. It works well because it is both self-serving and generous. In the past this impulse has often been obscured by events. The hindsight we now enjoy because of the massive historical changes of the past two centuries, especially of the last four decades, enable us to see that most social philosophers of the past have been wrong about the so called "cycles" of history. The very long term trend is, and always has been, towards increased individual liberty in the creation of human societies. The history of man as exemplified by the great nations of the Earth has proceeded in the following way:

Hunter Gatherer Bands - The dawn of civilization. To secure advantages man organizes.

Agricultural Feudalism - Strong centralized control. Exploitation leads to revolution.

Industrial Socialism - Capital is generated. Exploitation leads to further revolution.

Democracy with Capitalism - Populace becomes more developed. Libertarian refinements persist.

Libertarian Capitalist Republic - Highest potential for human societies (30).

The premise here is that the highest evolutionary destiny for all living things can best be manifested by continuing peace and prosperity in the human sector, and that all moral people want this. Most believe that such a condition, sustained on a worldwide basis, is an impossible utopian ideal. It is, however, only the current activities of governments which keep us from this very natural and easily attainable condition.

3. The innate love of liberty and the concession of this to others is the "Basic Libertarian Impulse". Depending upon the degree of spiritual development, the individual either manifests this or does not. We know that it is unjust to unnecessarily kill, assault, coerce, rob, defraud, slander, or otherwise encroach upon any living creature. Calling these acts by other names and programming an ignorant majority to agree that they are necessary or permissible does not change their true nature. To do evil is to trespass unnecessarily upon the liberty of any living organism. Historically this principle has been called the Golden Rule.

4. In human affairs we accept the premise that it is desirable for people to reach their natural level of prosperity and development through their own volition while living in peace and harmony with each other. A human being is an creature which comes into this world with no rights owed him and no obligation incumbent upon him, except the natural right to absolute individual liberty, and since he is not alone on the planet, the logical obligation of reciprocity in this towards others. He need only concede to others the same liberty he demands for himself, because this is absolutely all that is necessary for continuing harmony on Earth. The one human responsibility is simply to never make unjust encroachment. The only legitimate function of government is to enforce this natural obligation of mankind. Any person or government attempting to impose any burden other than this upon the individual is guilty of criminal coercion and should be dealt with as a mortal enemy even if the oppression is sanctioned democratically.

5. The natural inspiration towards liberty for all when acted upon, essentially renders the individual metaphorically a "Warrior of Light". Along with the idea of voluntary self-sacrifice, this principle is the only thing of universal value to be found in any moral system or religion. All other elements simply reflect peripheral things, sometimes of great archetypal value, but only to the particular cultural groups among whom they occur. The flip side of this is of course, that any individual who works against individual liberty is to that extent a "Slave of Darkness".

6. Individual liberty is the innate right to be free of unjust encroachment from others. It doesn't matter if the others outnumber us, are organized, and use euphemistic terminology. There are two types of unjust encroachment against individual liberty - illegal and legal. From a Libertarian standpoint both are equally as criminal. Un-Libertarian elements will allege that legal crimes are not unjust because they are determined to be necessary by a majority opinion and that this should supersede any objective measure of workability and rightness. People who don't want to be free are cowards. People who don't want others to be free are criminals. The majority of people on Earth always have been and still are, both. There is no reason for decent people to compromise about this. All of the crimes legally committed by collectivist governments are destructive to society. When they are continually perpetrated in defiance of known superior alternatives, they must also be effectively regarded as treason.

7. Every problem in every society on Earth can be traced back to a point where someone in government decides to sacrifice individual liberty for some other goal. Like any breech of natural law this produces a distortion. One compromise seems to justify another and soon the cause and effect relationships become obscured by time and complexity. The achievement of harmony on Earth simply involves eliminating the complex of false dependencies that have arisen because of these past mistakes.

8. Collectivist propaganda notwithstanding, interdisciplinary studies have objectively established that the Libertarian Capitalist Republic is the only political-economic system which has complete internal coherence and long term workability. This is because it is the only system which works in harmony with natural principles rather than against them. It is the one and only form of institutionalized human action which favors the ongoing evolutionary expression of all living things.

9. The highest aspiration of humanity is that the evolutionary destiny of life throughout the universe be unimpeded. The Quest here on Earth is the elimination of all deterrents to evolutionary expression through the actualization of a true World Libertarian Order, right now. This does not mean one nation or one world government, but a worldwide aggregation of separate Libertarian Republics, not completely attaining to this exalted manifestation simultaneously, but eventually. Take heart and reject all the post-Apocalyptic scenarios. Utopia is possible. Libertarian Nationalist Revolution is well underway and will ultimately produce this ideal condition everywhere in the world through the inspired participation of intelligent people everywhere.

How this will occur is far beyond the scope of this volume, but is clearly summarized at some length in another book "Libertarian Nationalist Revolution" described below.

7. The Triumph of Evolution

New World Order: Seek and Destroy

Countdown to Globalization

When the main threats to individual liberty
center around the impending loss of national
sovereignty and the destruction of indigenous
races and culture, then nationalism, by any
means necessary, including war, becomes the
first principle.

Sleepwalkers of the World, wake up! Forget your limp-wristed religious fantasies, alcohol intoxication, drug dreams, idiot ballgames, and virtual-reality heroism. Stand up on your feet like real men and women, think about the future, and show some proper adult seriousness for once in your pathetic lives. The eleventh hour is past. World Libertarianism is now the only alternative to globalist oppression. All the nonsense you think matters is of no importance, and never has been. If you feel insulted by all this, then you are one who needs to read further. If not, you are probably smiling at this moment, as we are.

Fully educated people know that a multiplicity of free sovereign nations competing in a free world market has a natural workability superior to any form of world government, and can be less easily subverted to collectivism. World problems will not be solved through ignorance. Join us grownups in the Twenty First Century. Read further, so that when the smoke clears you will be worthy to smile along with us, as one who has participated in the throwing down of globalist tyranny. Embark upon the Greatest of All Quests: Liberty Triumphant and Eternal!

Freethinker

Knowledge and Belief

Preliminary Principles:

1. An actuality is a pure state of existence apart from the perception of it by any living organism.

2. A reality is the accurate perception of an actuality by any healthy living organism. This will be qualified to some extent by previous experience and by the perceptual apparatus of the organism.

3. A fact is the conceptual representative of a reality.

4. Facts are the building blocks of correct thinking.

5. Logic is the process of correct thinking, the natural method used to arrange the building blocks provided by facts.

6. Knowledge is the correct natural correlation of facts by means of logic, the finished structure.

7. Truth is the broad and meaningful apprehension of knowledge.

8. Wisdom is the loving and just reaction to truth.

9. A belief, in the pure sense, is an attempt to extrapolate beyond what is known.

10. The amount and strength of an individual's beliefs is inversely proportional to the amount of his knowledge.

11. Philosophy is used to create a feeling of personal integrity and wholeness by attempting to extrapolate beyond available facts. Correct reactions are based upon facts, not upon philosophy.

12. Most philosophy is merely the "explanation" that people lacking facts offer to justify their own particular emotional reaction to their environment. The only worthwhile philosophy is a comprehensive overview of all available factual data fused by love, heroic idealism, good moral character, and courage. This involves an eclectic approach to the attainment of wisdom, not a slavish adherence to "isms" of any kind, including the fashionable zeitgeist of well entrenched science.

Viable Spirituality

In this context, spirituality is differentiated from religion, because it represents something much larger. It includes all of a person's values as these are reflected by actual conduct over the course of a lifetime, rather than by the mere parroting of religious doctrine, often with partial adherence. A person's spirituality comprises everything in life. This may include religious activity or no religion at all.

A truly viable spirituality must have perfect integrity between three basic components:

~ an intellectual premise consistent with all known science and which grows along with science

~ a moral premise reflecting absolute Libertarian reciprocity. This means no unjust encroachment against any creature or the environment to the detriment of any living thing. This also means absolutely no tolerance of such encroachment from others.

~ a source, not of dogmatic belief, but of archetypal inspiration, grounded in one's own ancestral mythology.

Absolute separation of church and state is impossible because what the people are spiritually determines the type of government they will create or condone. Wherever we find the institutionalized lassitude of cowardly religion, we also find Socialism or some other unworkable form of collectivist government. We can't expect an inferior prevailing spirituality to result in a superior way of running society.

Evolutionary Spirituality

This is spirituality based upon evolutionary principles. It will evolve as society grows in understanding. It will transform the world by inspiring people, first to cast off the chains which bind them individually, then to democratically throw down unjust governments everywhere. This is inevitable, but will happen only very slowly through ongoing education about what works and what does not.

Those who claim that human salvation can only be attained through obeisance to some popular savior or another are lying, although they usually believe they are not. The confusion and divisiveness these spiritual monopolists create with their bigoted power-hungry scheming is the one of the most destructive forces on Earth.

Salvation comes though living in accordance with truth and though the practice of righteous Libertarian principles.

Simple human decency is what will save the world, not mindless belief in tepid, limp-wristed mythologies.

Popular religions more or less advocate a morality based upon non encroachment against others. This is a good start, but he reason these religions have not saved the world is because they mix this moral truth with mythological falsehood and insist upon absolute literal belief. People see through much of the falsehood and then wrongly reject much of the moral truth along with the falsehood.

Personal Obstacles to World Liberty

There are many falsehoods that seem to give fulfillment in the short term, but which deliver absolutely nothing in the long term. They are only fantasy, and a complete waste of time. Dealing with reality is far more exciting, and produces constructive change, because attention isn't diverted away from things that really matter. The main problem areas are:

1. False Beliefs
One need not embrace ancient fairy tales as literal truth in order to live a moral life. This is not to say that good principles cannot be illustrated with traditional stories and allegories. We simply need to sort out what is real.

2. Non-Libertarian Economics and Politics
Collectivist systems simply do not work. All fully educated people know this. False systems include Communism, Socialism, Democratic Socialism, Social Democracy, Democracy not constituted upon Libertarian principles, Fascism, Populism, Theocracy, etc.

3. Controlled Media
Most easy-to-get information is controlled by those in service to globalist bankers and their subverted government lackeys, re-elected decade after decade by an ignorant apathetic majority. One must seek further for hard knowledge upon which to make important decisions.

4. Drug Induced Euphoria

If you don't like environmental circumstances, change them with action, not merely your experience of them by ingesting chemicals. Alcohol is used to suppress the cerebral cortex in order to liberate the reptilian complex. It gives timid people false courage in business and romance. One need only affect a reconciliation between Jekyll and Hyde in this regard. Psychedelic drugs seem interesting because they give the user a strange mental life. Since all human societies are based almost entirely upon lies, reading books and learning truth will give a far stranger mental life without bad side effects.

5. Spectator Sports

Ball games are just arbitrary simian competitiveness, a wishy-washy substitute for those afraid to get involved in things that actually matter.

6. Virtual Reality

Millions today resemble those poor souls who freebase cocaine, sitting all day addicted to their own brain chemicals, playing video games which give false feelings of heroism in seeming to defat enemies that do not exist. Wake up! There are real enemies to defat.

7. Excess Entertainment

Novels, movies, and music are great, especially when they inspire heroic Libertarian ideals and action. The amount need only be in balance with the other demands of real life.

8. Unproductive Friendships

This involves not wasting time with fools who suffer from any of the above delusions. Friendship should be casual and light hearted, centered around the mutual love of causes, professions, or hobbies. Meddlesome busy-bodies who encourage validation of life strategies through consensus should be avoided. It is always best to keep one's own council in order to develop inner resources. If you need advice, seek what the greatest minds in history have said,

not some doped-up joker you knew in high school.
Remember what George Washington said,
"Have no intimacy with worthless men."

Libertarian Basics

The Basic Libertarian Premise: It is wrong to unjustly encroach upon any creature or the environment to the detriment of any living thing. Most people agree with this premise. The disagreement is over what constitutes encroachment.

The innate love of liberty and the concession of this to others is what we may call the Basic Libertarian Impulse. Depending upon the degree of spiritual development, the individual either manifests this or does not. We know that it is unjust to unnecessarily kill, assault, coerce, rob, defraud, slander, or otherwise encroach upon any living creature. Calling these acts by other names and programming an ignorant majority to agree that they are necessary or permissible does not change their nature. To do evil is to trespass unnecessarily upon the liberty of any living organism. Historically this principle has been called the Golden Rule.

In human affairs we accept the premise that is desirable for people to reach their natural level of prosperity and development through their own volition while living in peace and harmony with each other. A human being is an creature which comes into this world with no rights owed him and no obligation incumbent upon him, except the natural right to absolute individual liberty, and since he is not alone on the planet, the logical obligation of reciprocity in this towards others. He need only concede to others the same liberty that he demands for himself, because this is absolutely all that is necessary for continuing harmony on Earth. The one human responsibility then is simply to never make unjust encroachment. The only legitimate function of government is to enforce this natural obligation of humanity. Any person or government attempting to impose any burden other than this upon the individual is guilty of criminal coercion and should be regarded as a mortal enemy even if the oppression is sanctioned democratically.

Liberty is the natural right of every living organism to manifest justly as an unimpeded participant in evolutionary destiny. This manifestation, to be just, must not unnecessarily interfere with the evolutionary expression of any other living organism.

Every problem in every society on Earth can be traced back to a point where someone in government decides to sacrifice individual liberty for some other goal. Like any breech of natural law this produces a distortion. One compromise seems to justify another and soon the cause and effect relationships become obscured by time and complexity. The achievement of harmony on Earth simply involves eliminating the complex of false dependencies that have arisen because of these past mistakes.

Individual liberty is the innate right to be free of unjust encroachment from others. It doesn't matter if the others outnumber us, are organized, and use euphemistic terminology. There are two types of unjust encroachment against individual liberty, illegal and legal. From a Libertarian standpoint both are equally as criminal. Un-Libertarian elements will allege that legal crimes are not unjust because they are determined to be necessary by a majority opinion and that this should supersede any objective measure of workability and rightness. There is no reason for decent people to compromise about this. People who don't want to be free are cowards. People who keep others from being free are criminals. The majority of people on Earth have always been and still are both.

People who have little regard for individual liberty will think that anyone who questions their morality or basic understanding in this regard is being terribly unreasonable. For those, however, who have a heroic vision for future societies, liberty is not a question to be begged, but a moral absolute. It is not negotiable or compromisible. Its value is not merely theoretical or just a "matter of opinion". The Libertarian position is the only viewpoint which is notaunreasonable. The people who now oppose individual liberty eventually will be vanquished. Natural order will prevail. The first line of attack is education.

Libertarianism is not power hungry politics, but the structuring of human affairs in accordance with natural law. Anarcho-Capitalism is not chaos, but the one and only system of economics which implements natural order.

The individual has the natural right to live in a free society, failing this, to live in liberty within any society in which he may find himself, regardless of the "consequences" to anyone else. Absolute individual liberty is the one and only thing worth fighting for. The truly Libertarian position is superior to all others, intellectually and morally. There are few however, who really understand or practice Libertarian ideals. No existing government and very few people will knowingly allow complete liberty to anyone if it seems to suit their purpose to do otherwise. If the individual wants liberty, he has to reach out and take it at any cost, must guard it jealously, and to keep it must be willing to fight even unto death.

No matter what else man accomplishes, if he does not immediately deal with the problem of increasing population, nothing else he does will matter. Couples who have more than two children make direct encroachment against all other creatures on this planet. The ideal population level on Earth was passed hundreds of years ago, if by the word "ideal" we mean a level consistent with concepts like individual self-actualization and opulent joy in living, rather than mere subsistence in anguished mediocrity. Evolutionary destiny is served through qualitation, not quantification. Maximum joyful manifestation for small numbers is superior to minimum meager manifestation for vast suffering multitudes. We are not imbued with life merely to endure it.

Trying to make the world Libertarian through writing is like holding a message in a bottle while standing upon a high precipice overlooking the vast expanse of limitless ocean. You are at the brink of heroic destiny, but casting your message into the sea of fate. The message is a knowledge of natural principles, which if generally acted upon by mankind, will ensure worldwide prosperity and peace, the unimpeded evolutionary expression of all living things. You wonder if the message will ever be read by anyone. Your only certainty is that little perceptible change will result in your own lifetime, because there are few who would comprehend the message even if everyone did read it. You wonder if the message will be preserved long enough to make any difference at all or whether the life you have spent has simply been wasted. Then you wonder if anything matters at all. Then you contemplate the other things that you might have done with your life and you realize that there is nothing, nor could there ever be anything, more excellent than this: The Greatest of All Quests.

Essential Facts for World Liberty

"Let me issue and control a nation's money and I care not who writes the laws." ~ Mayer Amsche Rothschild ~

"If the American people ever allow private banks to control the issue of their currency, first by inflation, then by deflation, the banks...will deprive the people of all property until their children wake up homeless on the continent their fathers conquered.... The issuing power should be taken from the banks and restored to the people, to whom it properly belongs." ~ Thomas Jefferson ~

"We are on the verge of a global transformation. All we need is the right major crisis and the nations will accept the New World Order." ~ David Rockefeller ~

Weed Out Falsehood

We are tired of crackpots who claim that predatory globalist bankers are aliens from outer space or another dimension. It seems more likely that those who say these things are working to engineer popular mistrust of the resistance. Additionally, since it is primarily events of the past two centuries which affect us now, the endless attempts to trace all this back to ancient Egypt or reconcile it with ancient prophesies, only complicates and distracts from the real issue at hand, the upcoming triumph of tyranny on Earth.

What follows is not "conspiracy theory" but well documented fact. It is not easy to see against the complex background of world affairs. Because of independent media, public awareness of these matters has been increasing in past months. It's a good idea to print out or get hardcopy when appropriate, because very resourceful people are trying hard to stop Internet access to information about these matters.

The New World Order

The predatory globalists are international bankers, not extraterrestrials, but they love only gold, and in their sick insatiable greed, rob all of humanity of the natural right to liberty, earned prosperity, and peace. They do this by manipulating currencies through privately owned central banks like the Federal Reserve Bank of the United States, and with the help of subverted politicians, engineer wars and economic upheaval so that they can lend money to governments for military mobilization and otherwise unnecessary social programs. This is what is meant by "Welfare-Warfare Economies." There is a large detailed body of historical fact about how they have done this for the past two hundred years.

Internationally these people are guilty of crimes against humanity on a scale greater than anyone in all of human history. In their individual countries they are, at very least, guilty of treason. They are allowed to continue in this only because of public ignorance. We can bring these enemies of all human potential to justice with legal precedents like those enacted at Nuremberg, but before an international tribunal can be convened and indictments issued, there must be increased public demand. Liberty-loving people need to learn about these matters and pass it on to others. This in turn must lead to activism: resolutions and petitions by business and civic groups to international organizations, senators, and congressmen.

At this time in history, there is no greater responsibility, and no higher calling. Taking refuge in endless popular modes of escapist delusion will not ensure the future of life and liberty on this planet. Even if adults no longer care about their own futures, they should get involved at least for the sake of the children. In the voting booth, there is merely the illusion of a choice between two NWO puppets. The only real choice is between World Libertarianism and ongoing "two-party" elections geared to globalist tyranny.

The US National debt is nearly twenty trillion dollars, every penny created by fiat counterfeiting. Nobody earned this money. Other nations will follow the US example when it works. All America needs to do is nationalize the Federal Reserve, repudiate the national debt, and demand reparations from the creditors for the amount already swindled from the American people.

Globalism and the Federal Reserve

The Federal Reserve Bank is the central bank of the United States. Out of thin air, by *fiat*, it creates paper currency and regulates the US money supply. Despite the misleading name, it is not part of the federal government. It is a private corporation owned by a cartel of international banking firms. The list of creditor-shareholders appears below.

The argument used by the bankers has always been that they can do a better job of managing things, so the government should borrow currency the bankers create by fiat, rather than create it on their own.

George Washington and other presidents kept the international bankers from taking over the issuing of currency for one hundred and twenty-four years. During this time there was a period of ninety years with total monetary stability, no rise in consumer prices at all. The only taxes were on real estate, tobacco, and liquor, and this was during the time of greatest immigration and road building in all of human history.

It wasn't until 1913 that the increasing number of corrupt politicians and media bosses made a banker takeover possible. Both the Federal Reserve Act and the Internal Revenue Act were debated in Congress, but many say that neither was properly ratified, and for that reason are both unlawful. This, of course, is a moot point, since both are unnecessary, destructive, and based on deception. The banker friendly controlled media has been ever vigilant in making the American people believe that both are not only lawful, but worthwhile. The entire swindle hinges only upon mass public ignorance about the facts presented here.

To *nationalize* the Federal Reserve Bank is simply to return the power to issue and regulate currency to the people through their government, which will no longer have to borrow or pay back the money. The unnecessary federal income tax pays only the interest on the unnecessary national debt which is now at nearly twenty trillion dollars.

Most other nations have a central bank equivalent in function to the US Federal Reserve. The globalist regulating agency for the entire worldwide banking cartel is called the International Monetary Fund. To get things back in balance, countries everywhere need to nationalize their central banks, repudiate their national debts, and demand reparation for the amount already swindled by the globalist creditors, as a civil alternative to being put on trial for engineering every war and ruined economy over the past two hundred years, or being the beneficiary heirs thereto, all of which is easily provable from existing historical records.This should be followed by a return to currencies backed by durable commodity of intrinsic value, like gold or a mixed store of precious metals, the value of which is determined in world markets.

US Federal Reeve Bank Shareholders:
Rothschild of London, Berlin
Lazard Brothers Banks of Paris
Israel Moses Seif Banks of Italy
Warburg Amsterdam. Hamburg
Lehman Brothers of New York
Kuhn, Loeb Bank of New York
Goldman, Sachs of New York
Levi P. Morton of New York
Hanover Trust of New York

Thomas Jefferson
3rd President of the
United States

"I believe that banking institutions are more dangerous to our liberties than standing armies. If the American people ever allow private banks to control the issue of their currency, first by inflation, then by deflation, the banks and corporations that will grow up around [these banks] will deprive the people of all property until their childern wake-up homeless on the continent their fathers conquered. The issuing power [of currency] should be taken from the banks and restored to the people, to whom it properly belongs."

Alternative to Globalist Tyranny

Most of the big problems on Earth are caused by parasitic international bankers who, via central banks, manipulate currencies, and with the help of subverted politicians, engineer wars and economic upheaval so that they can lend money to governments for military mobilization and otherwise unnecessary social programs.

The Shadow Government / New World Order agenda, called Globalization, is merely Totalitarian Socialism with One World Government, giving absolute monopoly to predatory bankers. Any reputable economist will tell you that what the World Libertarian Order proposes will result in ongoing liberty, prosperity, and peace for all people on Earth. Saving the world is a big job, but it's the best, and involves the making right of all past mistakes, not mere adaptation to the aftermath. No single group, governmental body, or army is expected to affect this entire program. Those who like the future vision depicted, should simply do whatever they can towards the desired end.

Without the Globalists, we will have a future for the entire world, free of economic upheaval, war, and pollution, with no encroachment on any living thing, indigenous peoples enjoying strict population control, race and culture preservation, absolute individual liberty, prosperity, and peace as separate sovereign nations competing in a free world market.

Following is a ten point program, which must be implemented to liberate societies so that natural order can prevail. The best approach will vary from one country to another because of what has occurred in the past, but the variations involve only short term emphasis and sequence, not policy or principles. The time frame for phasing in any particular policy must be of sufficient duration for smooth transition tp minimize any short term bad effects upon individuals or economies. What follows is a summary to establish proper sequencing. The reader will find some recap of previous material.

Ten Steps to Libertarian Nationalist Revolution

1. Revolution

Politicians:

All at the same time, stand up and be courageous. Show some integrity. Stop serving the New World Order parasites. Accept the premise that government is at best a necessary evil and that the less of it we have, the better. Please support all measures outlined herein.

Everybody Else:

Initiate revolution. Support all popular measures that are in a basic Libertarian direction, such as budget balancing. When there is no other choice, deal with gross encroachments against individual liberty covertly on an individual basis. Educate the upcoming generation at the grass roots level, about the sole workability of Libertarian principles so that un-Libertarian elements can finally be voted out of office everywhere. Maintain health, practice martial skills, and stay well-armed in case we get a chance to do it sooner. All this is the only difficult part. The rest is simple and could then be implemented quickly unless otherwise specified.

2. Banking and Trade

In every country, nationalize privately owned central banks, like the US Federal Reserve, repudiate the national debt, and demand reparations for the amount already swindled by the creditors, as a civil alternative to being put on trial for engineering every war and ruined economy over the past two hundred years, or being the beneficiary heirs thereto, all of which is easily provable from existing historical records. Return to currencies backed by durable commodity of intrinsic value, like gold or a mixed store of precious metals, the value of which will be determined in the world marketplace.

Consumers worldwide will have total product choice. Goods offered in the free world market will be produced solely within each country by the citizens of that country, with no foreign ownership of business anywhere. Banks will lend only within their own countries. Once all nations are prospering, few will think it good practice to invest away from home, and imbalances will subside. Simultaneously, phase out all subsidies and unjust regulation of business, trade, financial transfers, and banks.

For any bank, including the central bank, to maintain less than a one hundred percent reserve at all times is simply dishonest. The new policies will correct things easily with mandatory disclosure to depositors about actual amounts held in reserve, and clear information on what it all means

3. Taxation

Eliminate wealth redistribution at gunpoint, aka taxes, and institute specific user fees and designated lotteries. This will not happen simultaneously in all areas of spending, but immediately wherever possible. From here forward unnecessary foreign adventures by governments will have to be paid for only by those who support them.

4. War

In this new scenario, war will fast become just an unhappy memory. The energies previously squandered in these conflicts will be channeled into undersea farming, renewable energy technology, space exploration, and interplanetary mining operations. Defense spending everywhere can be cut to a safe minimum, substituting standing military with a skeleton crew of officers for the coordination of a voluntary citizen militia adequate to any emergency. To this end, replace frivolous athletics in the schools with basic martial and survival training.

5. Socialism

End the artificial sustenance of non-viability. Gradually phase out social programs and entitlements as the improving economy and rate of employment makes this possible in each particular locality. This will be done slowly enough so that nobody will be hurt. How quickly this can happen, however, will be a great surprise to most people. Nobody will be hungry in a Libertarian society. There will be an emergency fund to alleviate desperation caused by unpredictable local catastrophe or incurred disability. This can be funded by designated lotteries at the federal and state level.

6. Crime

Deal intelligently with crime:

- Legalize victimless crimes involving consensual areas of human contract. Free all those confined for victimless crimes with a public apology, a little money to tide them over, and a list of realistic job offers.

- Recognize the true bad guys: rapists, human traffickers, kidnappers, child molesters, child and snuff porn filmmakers, arbitrary murderers, and serial killers. These people are irredeemable constitutional psychopaths who have made an unforgivable breech with humanity. For the safety and moral integrity of societies they must be put painlessly to death. Opponents of this should appreciate that one needn't be a rocket scientist to figure out that all it takes to avoid being executed for these terrible things is simply not to do them.

- Replace prisons with self-sustaining isolation communities, several square miles with agriculture, livestock, and small manufacturing. As economies improve, the inevitable one percent of humanity simply unable to support themselves can be offered permanent sustenance by private charity as per specified terms, such as voluntary sterilization. Any such individuals refusing this option will have to shift for themselves. If this causes them to make encroachment on anyone else's liberty, they will be placed in isolation communities.

7. Protectionism

As the distortions produced by hundreds of years of unnatural coercive government slowly begin to subside, cautiously phase out all unnecessary or unjust protectionist measures such as unnecessary safety regulations.

8. Education

Institute programs in schools to teach children about what went wrong in the past and how Libertarian policies have improved everything. Explain the manipulative relationship that previously existed between international finance and politicians. Supplement this with rigorous teaching about devolving humanity, racial preservation, excess birth rates, birth control, disease, and all individual classes of drugs. Make understanding of all this requisite for promotion. Teach the truth for forty years before eliminating public education.

9. Adjustments

Make all adjustments associated with simple Libertarian decency and smart living.
Examples include:

- Stop unnecessary environmental pollution as soon as this is viable. No pay-offs for ten year "feasibility" studies or twenty year "implementation" programs. Just stop it.

- Institute requirements in livestock production, zoo administration, and pet ownership based upon humane, free-range, hormone/drug-free models.

- Stop the cruel decadent down breeding of pets into evolutionary non-viability. Sterilize the existing animals. Ask yourself why little dogs are so nervous and angry that they bark viciously all day, every day. Would you not be angry if captors had done this to you?

- Give national park and forest lands back to the native populations from whom they were originally stolen. This with the provision that they continue to run the lands at a high standard, for the enjoyment of all. Current non-native employees can be offered life tenure or a new job.

- Overhaul medicine, stressing nutritional solutions, both therapeutic and preventive, as opposed to only pharmacological and surgical options. Eliminate the duplicitous role of the physician as both personal doctor and commission salesman for drug companies. Allow doctors to prescribe only within generic categories, the specific choice of drugs being left to the patients who select for themselves on the basis of price and manufacturer reputation.

- Respect the right of individuals to decide when their life is no longer viable. Establish regional centers where people can be put into cryonic suspension, or receive a lethal injection and be cremated.

10. Population and Race

Deal decisively with the issues of population and race:

- History shows that smaller numbers of people in any given place work best, so long as there are enough to defend the borders. For the land mass of Earth, the *ideal* population is 320 million people. This number was passed c 900 A.D. By the word *ideal* we mean a level consistent with vital self-actualization and opulent joy in living, rather than mere subsistence in anguished mediocrity. To this end, rigidly enforce a limit of two children per couple. More than two is an unjust encroachment upon others, like house burglary. World population will slowly decline to workable levels everywhere. The projected ideal numbers are as follows:

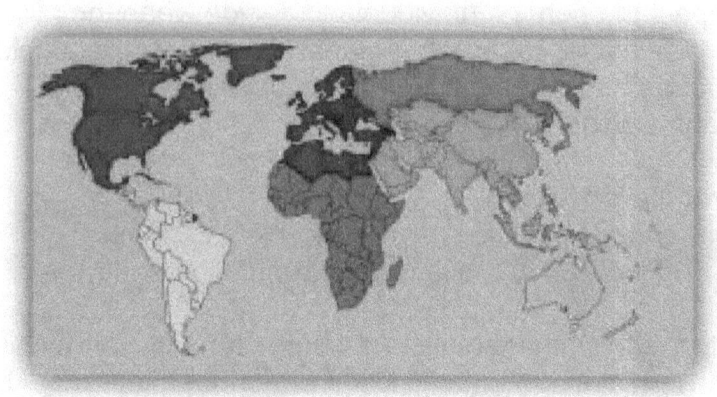

Canada, United States, Mexico
50,000,000

Central, South America
50,000,000

Greenland, Europe, Northern Africa
50,000,000

Southern Africa
50,000,000

Russia
50,000,000

Near, Middle East, Asia
50,000,000

Australia, New Zealand
20,000,000

- Workable societies must be based on natural principals. It is natural for people to feel most comfortable among those of their own race and ethnicity. In all of human history there has never been a multiracial or multicultural society which did not self-destruct because of the unnatural mixing. The New

World Order bankers, who work for totalitarian Socialism and world monoculture, want everyone to mix together, so they can lend money to national governments who must deal with all the resulting social problems.

- All people have the natural right to grow up among their own racial kinsmen. Resident racial outlanders are simply an unjust encroachment upon the personal liberty of indigenous peoples. To survive, we must emphasize race preservation and the prevention of global monoculture. Interracial marriage advocates are attempting to eliminate all existing races. They try to sound very interested in human variety, but their breed-up quick programs, long term, will completely obliterate human variety by making what are now separate races into one race. Variety is the spice of life. Imagine the tedium of universal sameness. The globalist bankers want to destroy race and culture. They know that one world government, giving global finance monopoly, will be more acceptable to people with no racial or ethnic identity.

- A common falsehood perpetrated by politicians in service to big business seeking cheap labor has been that ongoing immigration is necessary to keep industry alive. In actuality, business simply expands to accommodate any available work force. With worldwide prosperity, people will not flee their ancestral homelands.

- Note that third-world people usually favor globalization because it will allow them to prosper via social programs paid for by productive host populations. The predatory bankers know that countries with hordes of immigrant third-worlders, if the globalization question comes to a ballot referendum, will be far more likely to relinquish sovereignty.

- Close borders everywhere to immigrants of non-indigenous race. Anybody can leave, and a great many will begin to return home. Request voluntary sterilization of all who choose to remain in host countries, with special retirement programs for those who cooperate and adoption priorities for qualified couples within this category. No restrictions on travel. Tourists from now on will be able to enjoy the full undiluted potency of indigenous cultures everywhere.

- There will be new technology for determining constitutional psychopathy, even in the prenatal state, along with intrauterine diagnosis of fetal deformity, mental retardation, and genetic predisposition to sexual perversion. This will lead to the elimination of human non-viability everywhere on earth.

- Implement an equitable solution for the problems caused by a century of Socialism in unnaturally increasing the quantity, while undermining the quality, of people everywhere. There will be new foolproof brain-scan methods for determining intelligence. Use it to assess IQ in populations worldwide. Request voluntary sterilization of all those having an IQ of 94 or less, also with special retirement benefits and adoption priorities. Because higher moral conceptualization is a function of the cerebral cortex, these IQ adjustments, along with the elimination of constitutional psychopathy, will effectively spell an end to commonplace moral stupidity on this planet.

A Brighter Future

When all nations have attained free enterprise with global free trade, a basically Libertarian world will finally have been achieved. Evil will still exist on Earth at interpersonal levels, but it will no longer rule the day, nor will it never again be institutionalized by governments.

The Defeat of Globalism

Whenever possible it's best to conserve useful structure
created by the work of others, even when they are a
vanquished enemy. Besides the International Monetary
Fund, there are three Globalist institutions which we will
either reorganize to serve Libertarian Nationalist goals or
simply eliminate. They are, of course, EU, NATO, and the
UN. What follows are idealized situations in the future where
reason and voting produce the needed results. In real life it
will probably violent revolution. The alternative is to allow
money mad Globalists to destroy every race and culture on
earth. Imagine the tedium of universal sameness.

European Union

According to the leaders, the EU was formed: as a counter
balance to the United States as the only global super power,
out of the need for stability in Europe after the Second World
War, and as a product of economic agreements, all of which
still form its basis today.

In 1957, when the Treaty of Rome was signed to prevent
war in Europe. Fifty years later, with war still seeming far
away, the EU redefined itself as the leader against climate
change, but few Europeans could identify with greening up
the economy and lowering carbon emissions.

Hypothetically, an American Libertarian Nationalist leader
addresses the European Council:

"I have searched the web a good deal, and find that the EU
doesn't seem to have any clear idea of its own purpose.
Following is a edited comment from the EU Training Site:

"The EU says it wants to increase employment, and
prosperity, also to advance its innovation capabilities on a

global scale, but these are very vague ideas, and do not translate into a coherent narrative that one could call a "mission statement.

"EU leaders need to reach an agreement on what the European Union really stands for, and then summarize it, not in a ten-page European Council Conclusion, but in two or three sentences that are concrete and inspiring enough so that even a bus driver in Estonia will be able to understand...and maybe even support.

"How can European citizens be expected to support EU institutions and goals if they don't know what Brussels is struggling to achieve?

"Now I'm going to talk about what the EU actually does. The job of the European Union, as a federal government, strives to create and implement laws and regulations to dominate the member states. The leaders have mandated that the countries of the EU have uniform laws and policies for weights and measures, trade, labor, job choice, travel, immigration, rape, and politically correct speech. The EU enforces these laws.

"This enables the rights of sovereign member nations in these issues to be by-passed. If a member country wants to opt out, the EU threatens them with invasion by the EU army. The biggest EU initiative, in Globalist complicity with the UN, is to destroy the European race, culture, and civilization by flooding Europe with morally inferior peoples from whatever third world garbage pails they can get them."

There is a good deal of tension and murmuring in the auditorium as speaker continues:

"Anyone who has studied history knows that there has never been a multiracial society that did not self-destruct because of the forced unnatural mixing. There is absolutely no reason

71

for persisting in the current mode except the evil goal in serving the avarice of Globalist bankers.

"Indigenous populations in separate sovereign nations competing in a free world market is what works. The present course is not working, and can only lead to war. Europeans are waking up to all of this fast and will soon be gunning down racial outlanders in the streets for target practice on the way to work.

"Do the right thing while you still can. Stop further immigration, and send the resident outlanders home. Globalism is being defeated everywhere, but now there will be an infinite number of constructive projects for the bankers to finance. Isn't prosperity for all with lasting peace, better than prosperity for a few elitists, with endless warfare for the many? I know, what about population increase? Don't worry, we know how to reduce world numbers without mass starvation.

"The days of the EU are numbered. After just the pending withdrawals, it will be ludicrous to even call what's left European. If I worked here, I would be looking for the first available position with NATO or the UN. Help speed positive change. Please read about the Libertarian Nationalist program. It is the only arrangement that can lead to lasting worldwide peace and prosperity."

Many of the leaders look favorably impressed and the speaker is offered a great many handshakes on his way out to the cafeteria.

After two weeks of discussion; the European Union is officially dissolved. How could it be otherwise, with no viability for the future? Sovereignty for all the former member states will be preserved. It is agreed that he earlier national customs regulations will not be reinstated except along borders of indigenously non-European countries. The former currency exchanges are no longer necessary. With a little cashier technology, small business transactions can be for any amount, in any currency, anywhere.

NATO

When NATO was first set up, it was assumed that the US might be needed to help defend Western Europe from the Warsaw Pact. The original stated job of NATO was to be a defensive alliance to discourage war in Europe. Any member that is attacked can call on all the other members for help.

The only time this ever happened was September 11, 2001. The US asked NATO for help, and all the other nations responded.

NATO's mission statement as of 1999, did not envision any activity outside Europe and the North Atlantic. Today there is even division within Europe about how to handle the problems faced by these nations.

Anders Fogh Rasmussen, the NATO Secretary General has said,

"We will need a strategic concept that takes account of today's realities and tomorrow's challenges... The world has changed, the threats have changed, so has NATO."

 NATO's 28 member states hope that the future approach will help persuade an increasingly skeptical public in many European countries that the 60-year-old alliance remains relevant, decades after the end of the Cold War.

The task is harder at a time of economic crisis and shrinking defense budgets, analysts say.

Fogh Rasmussen's top priorities are expanding NATO's partnership with moderate nations in North Africa and the Middle East.

Hypothetically, an American Libertarian Nationalist leader addresses the North Atlantic Council:

"The North Atlantic Treaty Organization is not dead. Nothing likes to die, big international organizations included. Like anything living, it wants to be fed and continues to grow and extend itself.

"NATO was originally supposed to focus only on the common defense of its member states, and later threats like piracy, terrorism, and cyberattack. Today it gives itself free reign over the entire world. The excuses for NATO military actions appear limitless, along with the geographic arena in which it might take action.

"It made me sad to learn that NATO troops had amassed in Arizona to suppress the retaliation of American citizens to an expected economic collapse engineered by the Globalists. I remember a bumper sticker: Send our troops home. We need them to protect us from the government.

"Now that the EU is gone. and the UN will soon be reconstituted as an advisory body based upon Libertarian Nationalist principles, I and many like me in the US, feel that NATO should sever its now irrelevant allegiances to the Globalists and work with the newly emerging European nations to promote worldwide nationalism with the preservation of indigenous peoples and cultures everywhere. The new NATO agenda I suggest would be to review and get information to member nations about the superior options of worldwide nationalism, prosperity, and peace."

The speaker gets normal audience applause, followed by questions and answers.

After a few weeks, NATO is reconstituted to be completely in accord with Libertarian Nationalist policies and to continue as an advisory clearing house for the natters it used to act upon directly at the behest of the Globalists.

United Nations

Informal Mission Statement of the UN:

- To keep peace throughout the world

- To develop friendly relations among nations

- To help nations work together to improve the lives of poor people, to conquer hunger, disease and illiteracy, and to encourage respect for each other's rights and freedoms

-To be a center for harmonizing the actions of nations to achieve these goals.

Hypothetically, an American Libertarian Nationalist leader addresses the United Nations General Assembly:

"The UN originally had some very high sounding goals more or less associated with altruism. Unfortunately, it is today openly controlled by Globalist bankers. This is not surprising since it was they who treated it in the first place. The main functions of the UN now. of course, are the destruction of European race, culture, and civilization through the flooding of Europe with economic migrants disguised as refugees, and human trafficking in children for pedophilia.

"This shameful activity will not continue. Even the phony altruism will be opposed by all aware people from here on. Giving food to worthless layabouts only serves to suppress agriculture and to increase the numbers and moral lassitude of the recipients. We must guarantee all people an equal chance to succeed by their own volition. We must not guarantee them equal success apart from the quality of their participation in the process.

"The globalists are now irrelevant. The UN needs to wake up to this and purge your ranks of this foulness. Member nations will soon desert you if you don't get on board with what's happening now. Please study the literature I have provided on Libertarian Nationalism. With your talent and imagination, we can all live to see a lasting Golden Age for mankind on Earth."

The speaker sees a lot of scowls, and a lot of long overdue smiles in the assembly. This has been a very productive day indeed.

Soon the U.N. General Assembly votes that the United Nations is now to be an advisory body, committed to a Nationalist Libertarian future. All the UN Globalist functionaries have already been replaced with nationalists.

The new method of the UN will be to discuss and vote on the viability of worldwide policy suggestions, then pass on recommendations to national governments. For national leaders with good ideas, this will have the effect of being able to suggest legislation to everybody on earth, with the added legitimacy of the many-minds principle.

World Future Doctrine

World Revolution / World War III

"If the individual is born into an un-free society, he will have no legal rights corresponding to his natural ones, since these depend upon other people. He is endowed, however, in many cases, with the potential to be something more than merely a slave, and always with the choice of turning his will towards this end."

~ Dirk Aubrey Lokison ~

1. The world is in the worst trouble it has ever been in. There will be no help via supernatural intervention. We must fix it ourselves, and we can, but it requires good example through action, along with educational speech. Those who will not acknowledge any of this are part of the problem and should be treated accordingly,

2. If we don't reverse our increasing numbers immediately, nothing else we do will matter. We must have strict population control now, enforced with an iron fist if need be. In an overcrowded world, irresponsible couples who have more than two children are recklessly endangering everyone else and will not continue in this.

3. There are many races on Earth. They are not equal. The highest are, on the average, thirty-five IQ points above the lowest. All races, however, have equal potential, by their own long term own effort, but only if they are not interfered with by missionaries or those seeking colonial dominion. Unevolved people who have done nothing but hunt, gather, sing, and chant for the past twelve thousand years, of course, would just love to breed their way into developed, civilized societies. One is not being hateful or bigoted to recognize these irrefutable facts.

4. There is one superior race who wants total ownership and control of everything on Earth. They want everybody else's earned prosperity, and false credit for everybody else's achievements. These goals are clearly stated in their earliest writings penned thousands of years ago. All over the world, they infiltrate every area of human endeavor that can give them power over other people. They try to convince humanity that they deserve special advantage, and that they have been persecuted by everyone else, especially certain nations. Through monopolistic media control, they create false feelings of guilt by using selective emphasis, massive exaggeration, and outright lies. They seek to stifle decent by sponsoring legislation against what they label "hate speech" which is any speech that tells the truth about their activities and plans. They try to enforce the Doctrine of Political Correctness by every means possible. This is the body of principles debated in Russia after the Bolshevik Revolution, to be held dogmatically by both Socialists and Communists, a common core of unnatural non workability. In their scheme for world domination, their own country, Israel, is merely a home base to work from. or when necessary, retreat to. Once they are all in that country, without nuclear capability, their unjust adverse effect upon others will be neutralized.

5. Most world problems are caused by parasitic international bankers who, via privately owned central banks, like the US Federal Reserve, manipulate currencies, and with the help of

subverted politicians, engineer wars and economic upheaval so that they can lend money to governments for military activity and otherwise unnecessary social programs. We can rid ourselves of this problem by restoring the right to issue and control currency to the people. This will be accomplished by nationalizing all central banks worldwide. Reputable economists have developed workable models showing how this can be done without causing economic collapse.

6. In all of human history there has never been a multiracial or multicultural society which did not self-destruct because of the unnatural mixing. The predatory bankers exploit even the early stages of social decay, because they lend money to the national governments who must deal with the resulting problems.

7. The New World Order / Shadow Government wants most of the races of mankind to become extinct through forced intermixing. They know that world government, giving global finance monopoly, will be more acceptable to apathetic people with no racial or ethnic identity, living in a tedious gray landscape of universal sameness.

8. People have a natural right to grow up among their own racial kinsmen. Immigrant racial outlanders are an unjust encroachment upon the liberty of indigenous peoples. To survive, we must implement policies of race preservation to avert global monoculture everywhere.

9. To summarize, what works best is indigenous peoples enjoying strict population control, race and culture preservation, with absolute individual liberty, as separate sovereign nations competing eventually in a free world market. This will lead to a lasting peace and prosperity, an ongoing Golden Age for all of mankind. We have a right to implement this now, and to eliminate those who interfere.

10. People of all races need to look deep within themselves to isolate what is frivolous, and discard it. This is an excellent time in which to practice martial skills. Integrity will be gained by reading the facts and openly speaking the truth, about World War II, the last major conflict on earth. Inspiration can be gained by reading the biographies of men like Heinrich Himmler and Reinhard Heydrich to understand the tremendous obstacles they had to overcome

11. In a country where the government consists of subverted traitors in service to globalist bankers, there is no remedy at law. Those who ruin or imperil the lives of others for pleasure or profit, are combatants in an army of darkness, whether at the level of immigrant rapists or sponsoring globalist bankers. Moral people are empowered by the simple goodness within them to deal covertly with these enemies of all life. Such initiatives should be individual and never discussed with anyone, including spouse, parents, children, or friends. In the degenerate world of today, liberty and natural order will not be granted, so it must be seized. Soon we will all be involved in a worldwide revolution that will prevail by any means necessary.

Servants of oppression,
dig your graves!
Winter is coming.

New World Order

Short History

In 1773, Meier Amschel Rothschild met in Frankfurt, Germany with twelve of his most influential friends. He convinced them that by pooling their resources, they could rule the world. Rothschild soon found a man of incredible intelligence and ingenuity to head their organization, one Adam Weishaupt, a professor of Canon law. and Jesuit priest.

At the request of Rothschild and his friends, Weishaupt abandoned the Catholic Church, and created the secret Order of the Illuminati, on May 1, 1776. The objectives were, through currency manipulation via central bank control, and the cooperation of subverted politicians, to establish what they called a New World Order, which will seek:

Abolition of all ordered governments
Abolition of private property
Abolition of inheritance
Abolition of patriotism
Abolition of family
Abolition of religion
Creation of world government

Concern about the unelected power of international bankers arose as Congress prepared to extend a twenty-year charter with the Bank of the United States, a private central bank formed in 1816. Andrew Jackson vigorously opposed efforts to strengthen the grasp of any central bank over the U.S. He called private central bankers a "den of vipers" and in 1832 vetoed a bill to renew the charter.

Thomas Jefferson warned,

"If the American people ever allow private banks to control the issue of their currency, first by inflation and then by deflation, the banks and corporations that will grow up around them will deprive the people of all property until their children wake up homeless on the continent their fathers conquered."

The New World Order's power is nearly total today, with presidents, senators, congressmen, and mainstream media bosses among the subverted. The opposition to banker monopoly has shifted from those who govern to those who are governed, the people directly affected by the banker's totalitarian agenda.

Alexander Solzhenitsyn, in a speech at an AFL-CIO meeting, July 1975, spoke of a turning point where our hierarchy of values may waiver or collapse:

"The political crisis of today's world and the oncoming spiritual crisis, are occurring at the same time. It is our generation that will have to confront them."

World War II and Causes

Relatively Common Knowledge
Bankers and Subverted Politicians
"Lending Requires Spending"

Hebrew goals in early fiction unite them as a tribe.
Isaiah 60, 61 Covenant promise of Jehovah to the Israelites:
"Therefore, thy gates shall be open…that men may bring
unto thee the wealth of the Gentiles...For the nation and
kingdom that will not serve thee shall perish...Thou shalt also
suck the milk of the Gentiles...Ye shall eat the riches of the
Gentiles, and in their glory shall ye boast yourselves."

1917 Russian Revolution
Jewish Communists murder and confiscate the property of
twenty million Christians.

March 1919 First Communist Party Comintern Congress
Of the three hundred ninety-three delegates, all but
seventeen are Jews.

June 1919 Treaty of Versailles
There is no evidence that Germany started World War I, but
with the help of baker accomplice, Colonel Edward Mandell
House, Germany is blamed for the war, forced to

demilitarize, give up large territories, and pay huge reparations, so the bankers can lend the money to Germany. They know that the German effort to stop Communism will provide an excuse for another banker windfall in a Second World War.

In the wake of Versailles, roaming Communists are shooting German citizens in the streets. Adolf Hitler's SA Storm Troopers stop them. Josef Goebbels describes Communism as the "dictatorship of the inferior." Heinrich Himmler reaches the conclusion that the war against subhumans, because of the vast numbers, can never be completely won, but that good people must forever fight simply to hold the line.

Germany's huge reparations require nonstop fiat currency which causes massive hyperinflation. Gentiles must sell everything just to eat. Jews buy up large amounts of real estate in Germany. In Czechoslovakia they acquire eighty percent of the property. Germans grow very tired of Jewish Communists and bankers, and decide to find the Jews to a homeland of their own.

November 1938 Polish Jew, Herschel Grynszpan, enters the embassy in Paris to murder the German Ambassador, who is away on business. Instead, he shoots an assistant, Ernst Vom Rath. Germans want peace, and react angrily to this by breaking Jewish shopkeepers' windows. Goebbels has the SA encourage the activity. Finally the incident is stopped by the SS at Hitler's order, and is named the Night of Broken Glass, Kristallnacht.

September 1, 1939 a plebiscite has established that the resident population of Danzig, one of the territories lost at Versailles, wants to be re-annexed to their homeland, so Germany invades Poland.

September 3, 1939 Lord Halifax, the bankers' choice, has replaced Neville Chamberlain as the British peace negotiator. The bankers and paid-for politicians are now able to take England and France into a Second World War.

September 17, i939 Russia invades Poland, and on November 30. Finland.

In furtherance of Adolf Eichmann's plan for the creation of Israel, Germany wants to send a ship to Madagascar with thirty thousand Jews on board. With the outbreak of war, the plan is spoiled because of French shipping blockades. Reinhard Heydrich has been conducting a successful resettlement program for the Jews, but due to the war this too comes to a halt. Ultimately, Germany asks twenty-five different countries to take their Jews, but nobody will have them.

Germany negotiates for an end to the war, but Sir Winston Churchill, with a chess master's personal obsession to defeat Hitler, persuades the Allies to insist on unconditional surrender. For Germany, this would mean another Versailles, or worse, so the war drags on, and the Jews are shipped to work camps, mostly in Poland. The Allies succeed in bombing German supply lines. This leads to malnutrition and disease in the camps.

January 27, 1945 as the war winds down, the Russians liberate Auschwitz and other camps in Poland, but will not allow the press inside any of them for another five years, a year after the final verdicts at Nuremberg. When the camps are opened, there are lots of gas chambers that no one remembers seeing who visited the camps during the war.

In Berlin, invading Russian soldiers see flush toilets for the first time and think they are potato washing machines. They rape fifty thousand German women. Three million Germans are murdered after the official end of the war, two million civilians, mostly women, children and elderly, and one million prisoners of war. British historian Giles MacDonogh details how they are killed in cold blood, or confined and left to die of disease, cold, malnutrition, or starvation.

At the Nuremberg Trials, seventy-five percent of the staff lawyers are Jewish. Controlled media, friendly to Jewish banker goals, latch on to a careless estimate offered by a Vermont magistrate, that six million Jews were gassed in the camps. The higher the death toll, the bigger will be the loans for Germany's ongoing reparation payments to Israel.

American forensic doctors examine hundreds of bodies, but can't find even one that was gassed, most having died of typhus or starvation. Throughout the war the Red Cross, under the rules of the Geneva Convention, visited each of the camps once every two weeks. They say the total number of people who died in the camps is 271,301, including non-Jews. Based on census data, the World Almanac for 1940 gives the world's total Jewish population as 15,319,359. For 1949 it puts the number at 15,713,638.

For Auschwitz specifically, the plunderers claim a number of four million, but the top Jewish authority on Holocaust demographics, Gerald Reitlinger, says the number for Auschwitz is three hundred thousand. The banker media, however, hold fast against all the updated estimates, and always make a point of stressing the activity of atypical Germans like Dr. Joseph Mengele. Decades pass before we hear about Oskar Schindler.

February 2016. if anybody disputes the six million number, they are immediately labeled a "Holocaust denier." Banker friendly publishers have almost exclusive control of all mass media, and with tribal singularity of purpose using selective emphasis, have turned three generations of white Europeans into guilt ridden self-haters, blindly acquiescent in New World Order plans to ruin national economies, destroy European culture, and eliminate the white race with endless immigrant hordes of Third World savages. Any white person who will not proactively participate in the extermination of his own race, is labeled a "hater" or a "white supremacist." The next time a subverted globalist traitor invites you to a "conscious raising" Holocaust seminar, hang him for treason.

Jewish Globalist injustice in Europe today is legendary. The escalating plunder of the United States has taken the standard of living of the average American down forty percent since 2000. Financial aid from the US to Israel amounts to three thousand dollars a year for each Israeli family of four, in a time when Americans can't even pay their own mortgages. The aid to Israel is the reason for another banker delight, the continuing US war with Islamic nations, including the attack on the World Trade Center.

Fifty years ago, the US was the world's biggest creditor nation, now it is the biggest debtor nation. The US National debt is nearly twenty trillion dollars, every penny created by fiat. It's only controlled media disinformation that keeps Americans from knowing that the government can create its own currency, and tie its value to receipts for hours worked. We don't need to borrow from a Globalist banker cartel.

Living Organisms

Protozoa

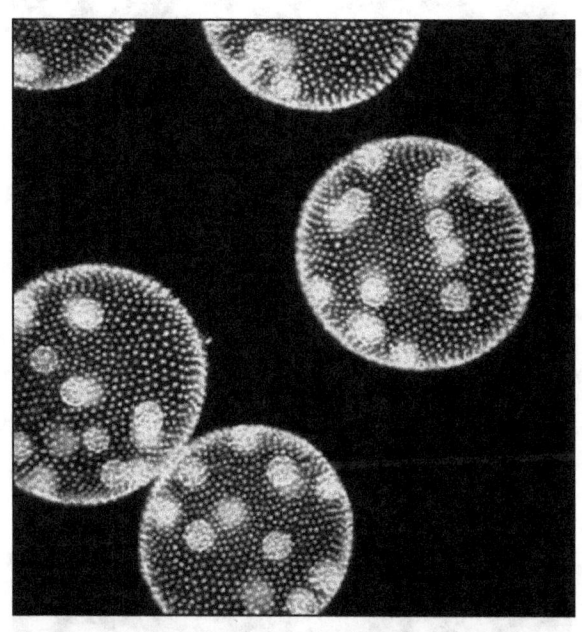

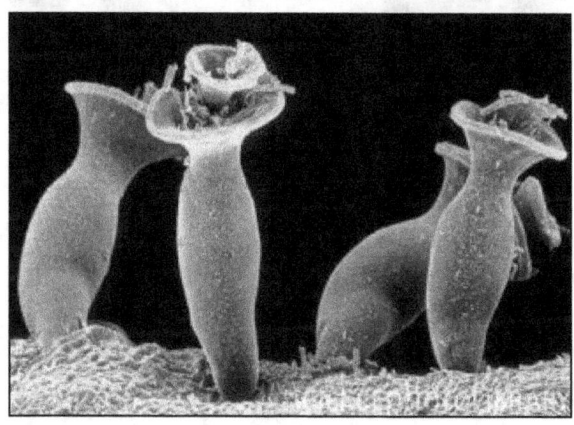

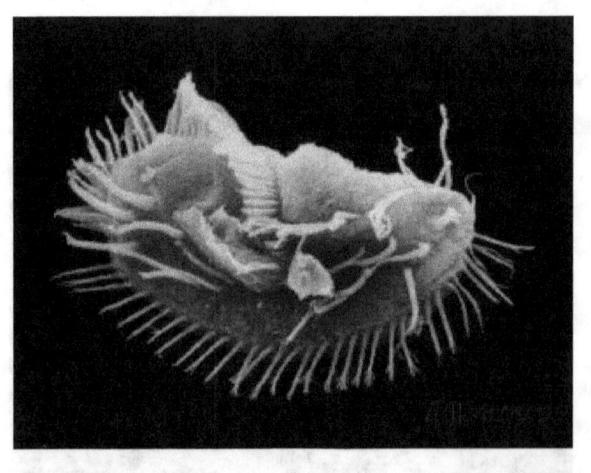

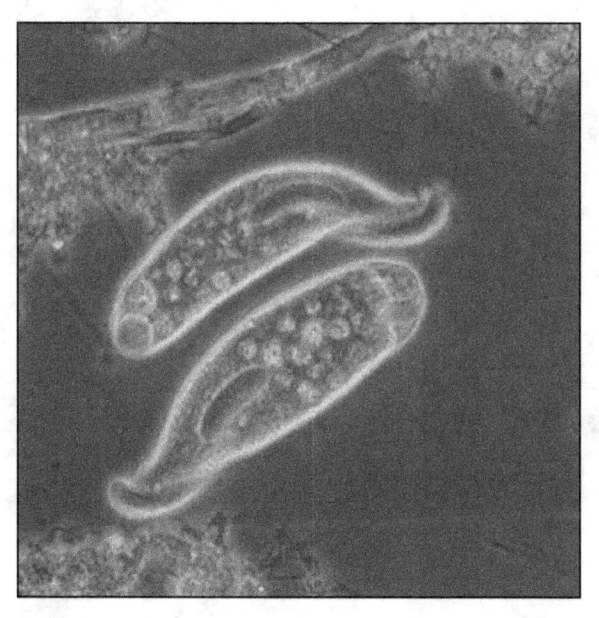

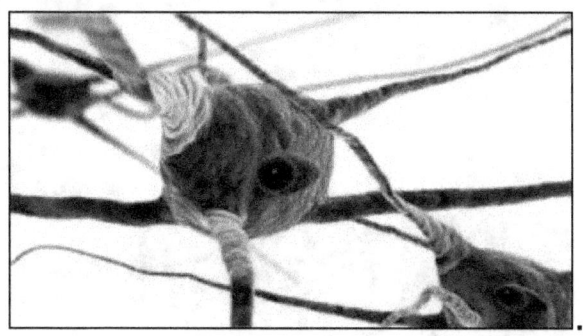

Before Dinosaurs

Sail-backed Reptiles (1/25)

Terrestrial Reptiles

Aquatic Reptiles

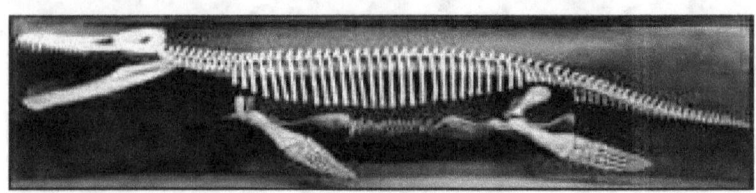

Mosasaurus

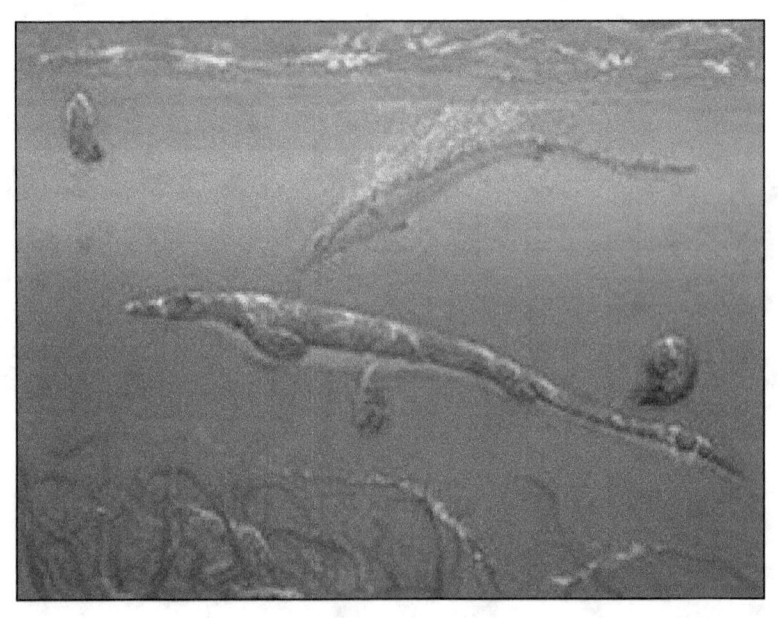

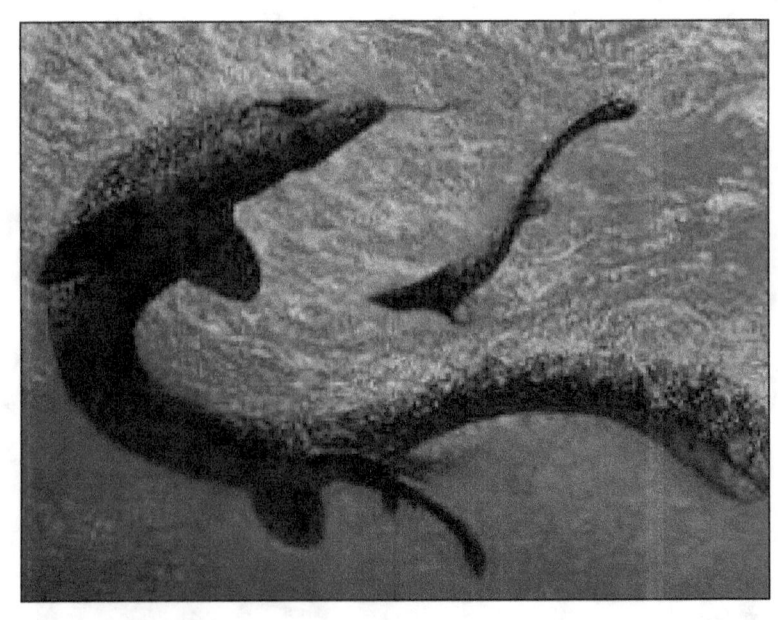

Aerial Reptiles

Prehistoric Mammals

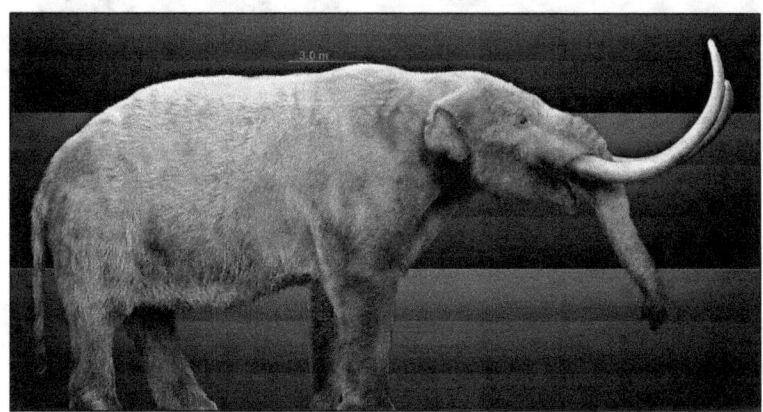

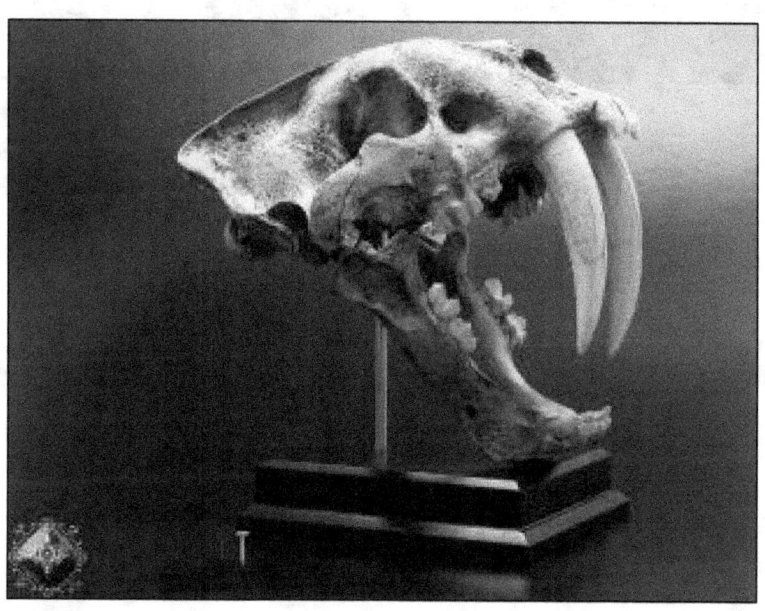

Modern Reptiles

Modern Mammals

Modern Birds

Modern Fish

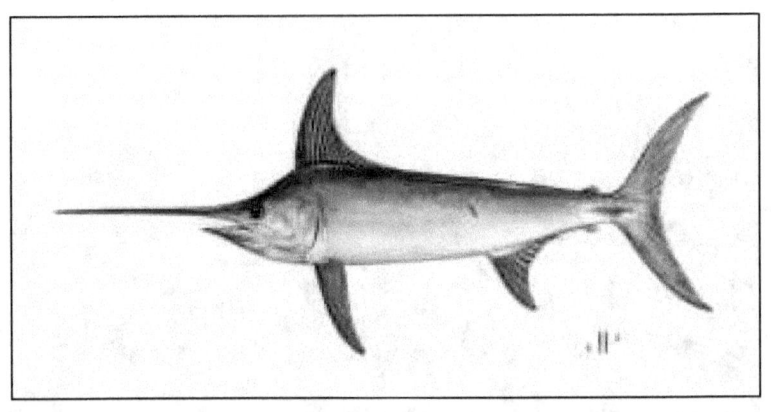

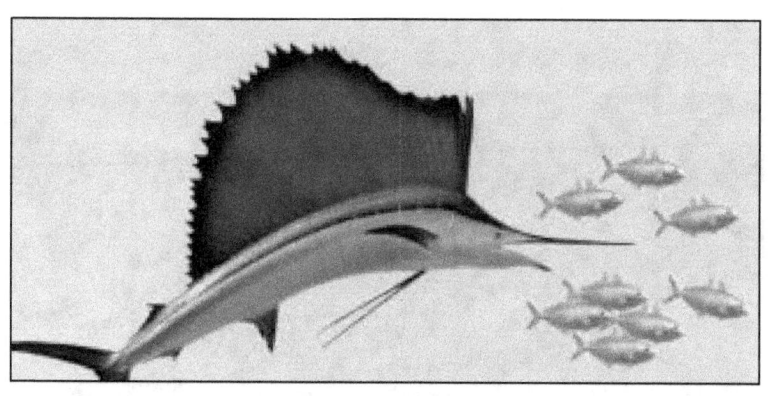

Modern Trees

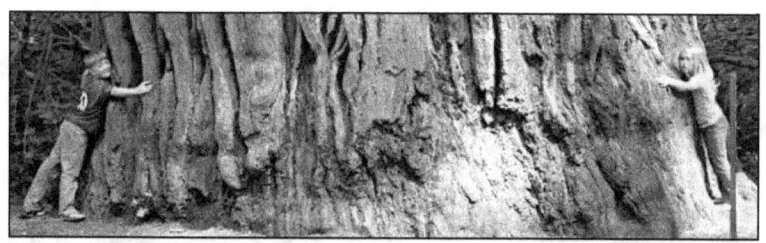

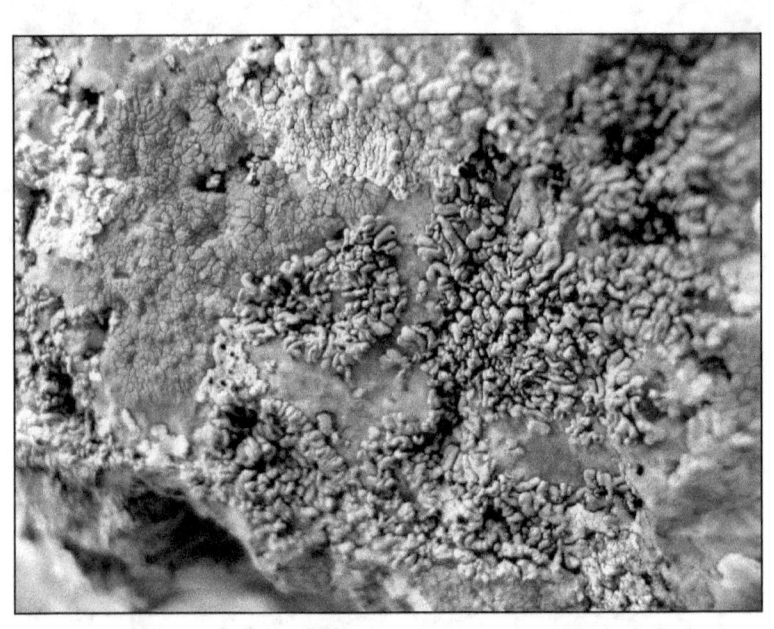

Bioluminescence

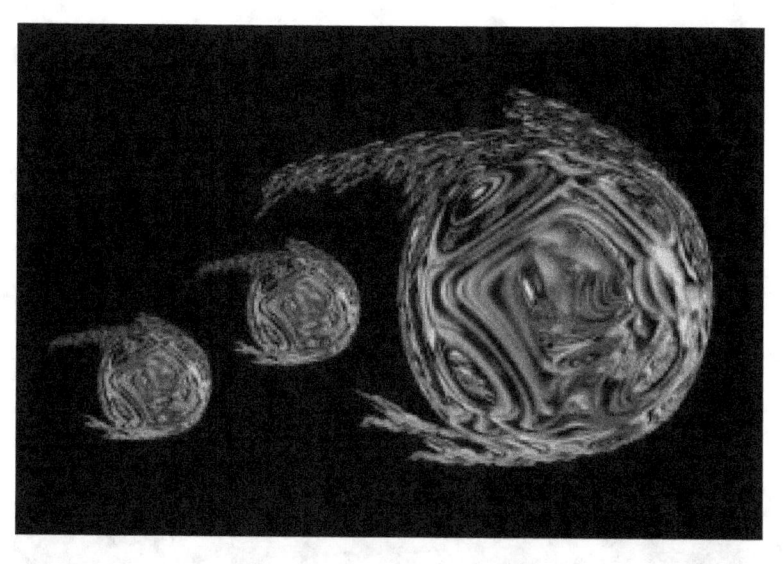

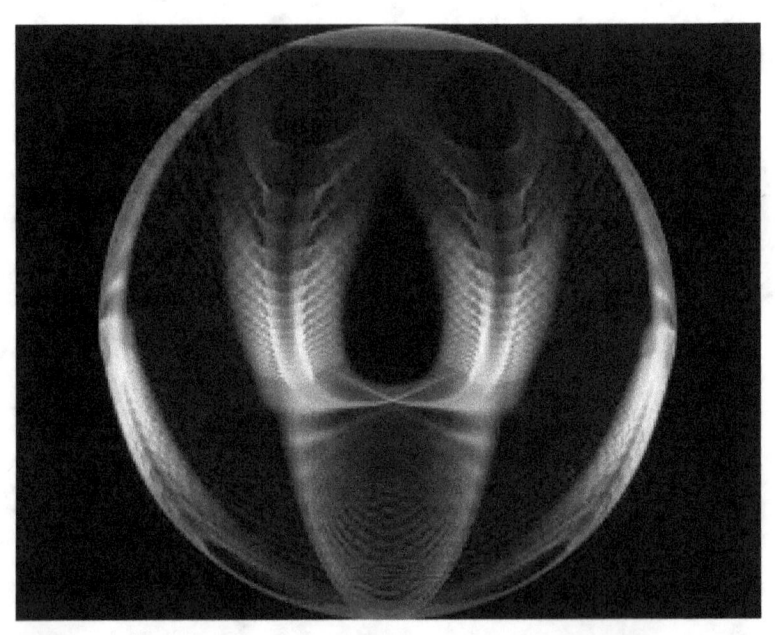

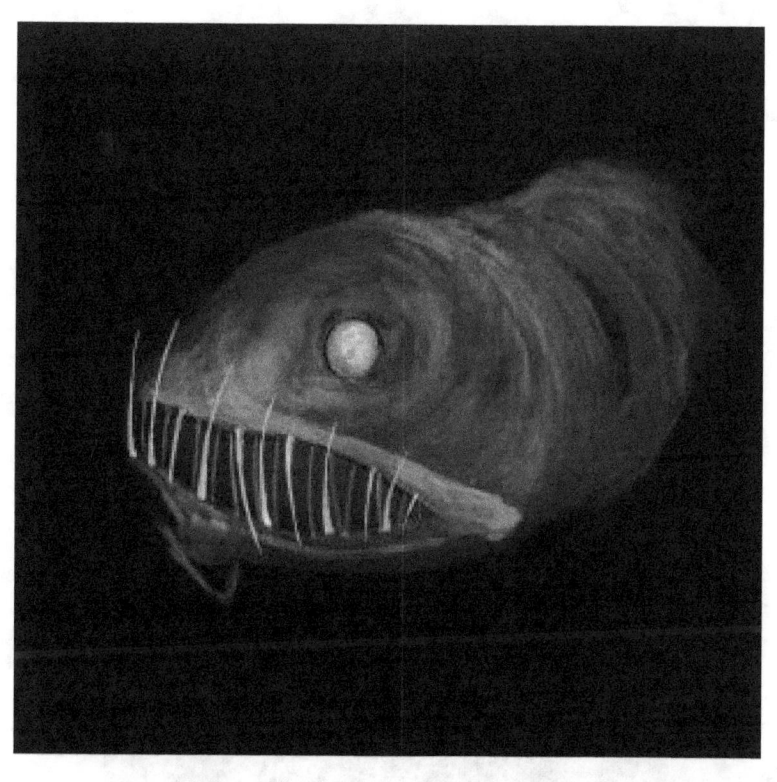

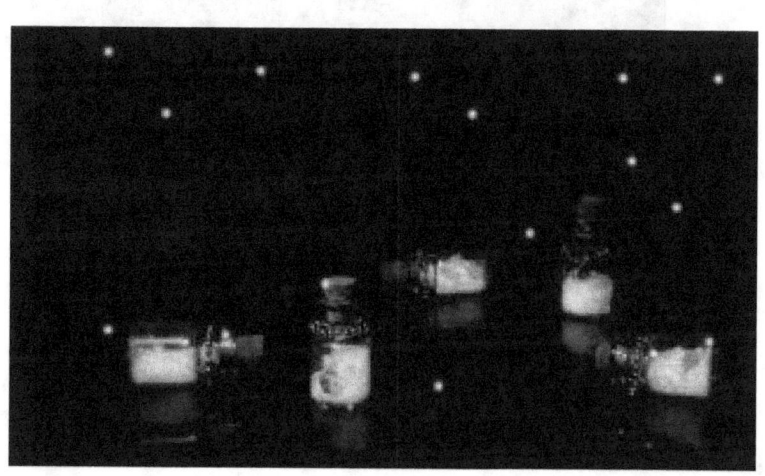

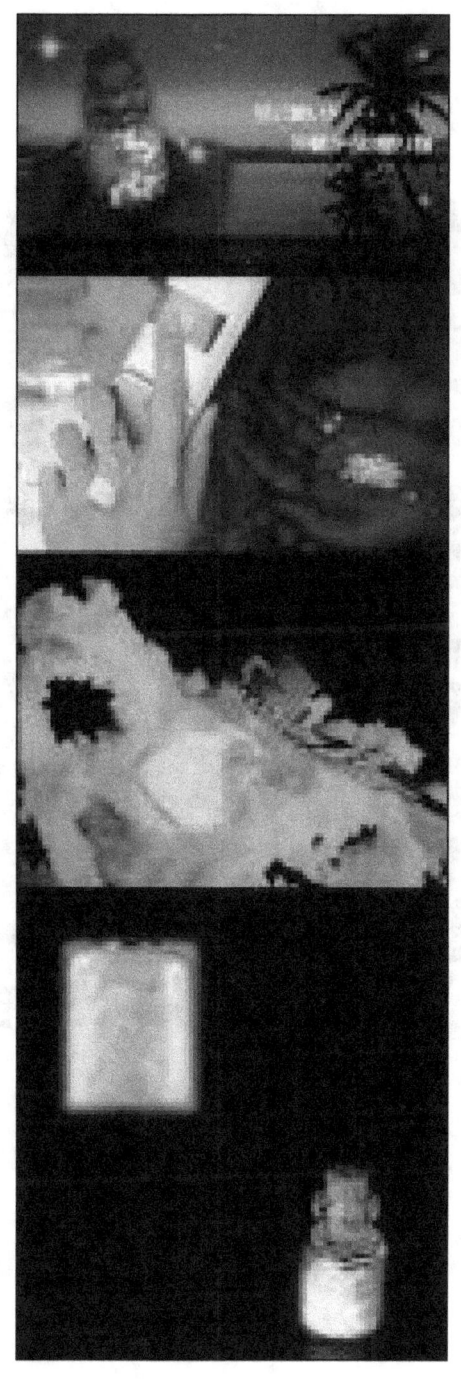

Cute or Unusual

Fairy Ring Mushroom

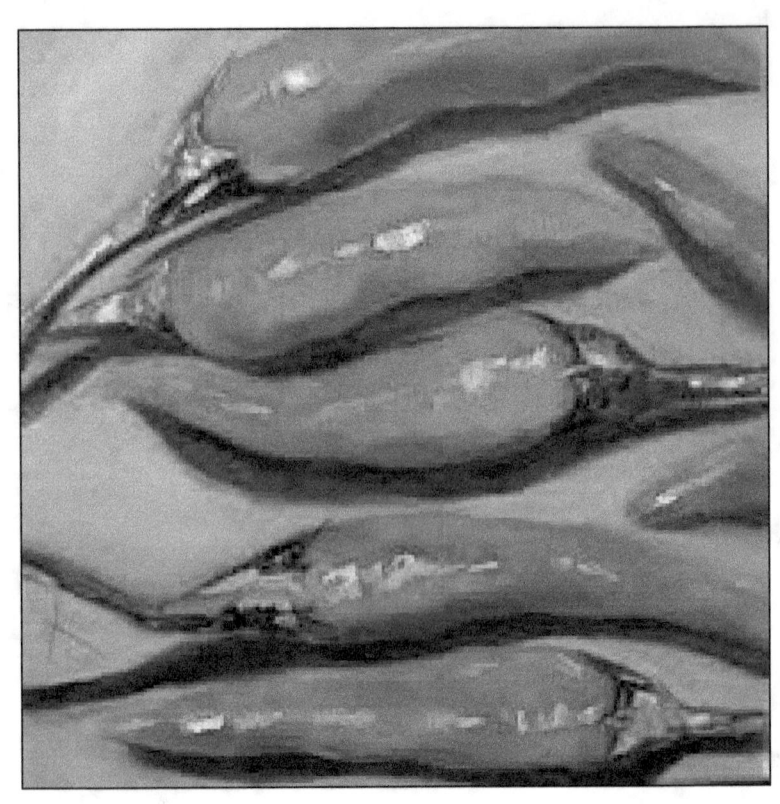

Geological Time and the History of Life
Illustrated by Animals & Plants of North & South America

ANCIENT LIFE PUBLISHING

Stop the Parasites

It is important to view these books and videos, because globalists control the mass media. They slant the news to destroy ethnic and cultural identity, so that host populations will accept one world government, giving their banker associates absolute financial monopoly. They do not use logical persuasion, but In a matter-of-fact way, suggest that the majority of people already believe in their goals. This is to make us feel that we will be out of step with current trends, and be disliked for not embracing the same viewpoints.

The subconscious mind is pre-lingual and cannot be influenced by words. Whenever possible, the media masters program us with pictures designed to elicit primal emotions. Even if we find out the truth from statistics later, the subconscious will still believe in the pictures.

We must rid ourselves of these hell-rotters once and for all. We cannot learn about superior alternatives to globalization until there are laws to protect societies against media monopoly. The fairest way is to require that the percentage of media ownership by any special interest group not exceed the percentage of that group in the national population. Who, but monopolists, would object to this? Read how things stand now, then ask yourself why any of this is tolerated:

Non-Fiction

New World Order: Seek and Destroy
from Viking Media Favorites
This compilation from many sources explains all you will
ever need to know to maximize your resistance to predatory
globalization.

None Dare Call It Conspiracy by Gary Allen
Riveting inside history of globalist bankers right from the
beginning. More compelling than the best of novels. Only
chumps, jokers, and sleepwalkers have not read this
one yet.

The Occult Technology of Power by Robert Eringer
Explanation of how the Shadow Government rules, written
as though by one of the globalist bankers to be
read posthumously by his son as instruction on how to wield
his newly inherited power.

Our Nordic Race by Richard Kelly Hoskins
Explains who the Nordic peoples are, how their civilizations
have been destroyed in the past, and urges future
preservation of the Nordic race and culture.

Why Civilizations Self Destruct by Elmer Pendell
Scholarly history of the way in which earlier societies fell into
decay as the entire world is doing now.

The Fulfillment of Evolutionary Destiny by Eric F. Magnuson
Explains how we can defeat globalist totalitarian socialism
with a far more workable worldwide Libertarian Free
Enterprise system.

Revolution: And How to Do It in a Modern Society
by Professor Kai Murros
Things are happening in Europe that should be happening elsewhere.

Holocaust: 120 Questions and Answers
by Charles E. Weber
From the Institute for Historical Review. One of many interesting contra-orthodox volumes refuting standard wartime disinformation.

Fiction

Hunter by Andrew MacDonald
This engrossing novel explains the truth about many world problems, including how to kill the everyday public enemies of your country covertly as a heroic citizen.

New World Order: The Final Solution by Roy C. Peterson
Exciting novel explains how to exterminate the growing legions of sub-humanity in massive numbers privately, but also how to legally establish world liberty, prosperity, and peace without killing anybody.
:

Eric F. Magnuson Short Biography

Eric Fenris Magnuson was born in Massachusetts. His parents were corporate business people. At Northeastern University, he studied science and English. Supporting himself as an assistant archaeologist, he amassed a library of over four thousand books and began a diverse program of private study. Moved by the need to create something that would outlive him, on February 12, 1983 he founded an activist organization, the World Libertarian Order. After a six year tour du ski. he moved to Lake Wildwood California, and at present continues his writings in Montreal, Quebec.

Fimbul Winter Books

Writings of Eric F Magnuson

Balanced Healthy Living / Absolute Individual Liberty /
Viable Evolutionary Spirituality

As director of the World Libertarian Order, I have worked for
peace and prosperity since the early 1980s. Most people
prefer fantasy to reality. Since my books deal only with
uncompromised truth, they are for the few, not the many.
I offer these writings for whatever good they may ultimately
accomplish in the world. They are all good quality glossy
paperbacks at a low price. To see them, visit your favorite
book vendor (e.g. Amazon, Barnes + Noble) and search
"Eric F Magnuson " under Books. You fill find independent
reviews and author descriptions.

~ Eric Fenris Magnuson ~

Evolutionary Psychology
Eric F. Magnuson

Evolution Family Reader
Eric F. Magnuson

World Libertarian Revolution
Eric F. Magnuson

New World Order
Just Say No!
Eric F. Magnuson

Mythology of the North
Eric F. Magnuson

Arcane Fraternal Orders
Eric F. Magnuson

The
Adventures
of
Eric F. Magnuson

Book I

The
Adventures
of
Eric F. Magnuson

Book II

www.ingramcontent.com/pod-product-compliance
Lightning Source LLC
Chambersburg PA
CBHW072303200526
45168CB00014B/256